D0906063

Technological and Social Change:
A Transdisciplinary Model

Technological and Social Change

a transdisciplinary model

Jacob Fried
PORTLAND STATE UNIVERSITY
Paul Molnar

Printed in the United States of America.
1 2 3 4 5 6 7 8 9 10

Library of Congress Cataloging in Publication Data

Fried, Jacob.
Technological and social change.

"A Petrocelli book."
Bibliography: p.
Includes index.
1. Technology—Social aspects. 2. Technology and civilization. 3. Technological innovations. 4. Social change. I. Molnar, Paul, joint author. II. Title.
T14.5.F73 301.24'3 78–11632
ISBN 0–89433–074–8

Dedicated to the memory of
my mother and father (J.F.)

contents

Part III: Theoretical Development

list of tables

preface and acknowledgments

Though this work is the product of two anthropologists, and we hope our colleagues in the social sciences will read it with interest, its major intent is to provide administrators, planners, and policy makers with a tool of analysis for treating extremely complex technological problems that are firmly imbedded in social and political processes. It is our aim to bring to the policy and management sciences, via this model, aspects of the classic holistic frame of reference that anthropology utilizes in its concept of culture, that is, a total design of technological, social, and cultural features. At the same time, we wished to avoid the generally leveled criticism that the "soft" social sciences utilize concepts and variables that are impressionistic and non-quantifiable, by inventing variables that can be given quantitative expression.

The model grew out of a long series of involvements with technological issues, beginning with an exploratory survey of the most complex technologies to be found in the Portland, Oregon area. This was undertaken in 1969 in order to determine in what different manners complex technologies involve social, managerial, and political aspects in their organization. Professors Morris E. Weitman and Milton Davis of the Psychology Department of Portland State University worked with us in an interdisciplinary venture which led to a publication which demonstrated that certain man-machine interfaces directly correlated with high rates of absenteeism (Fried, Weitman, Davis, 1972).

A graduate seminar grew out of these experiences in which we sought to extend our analysis of technological and social spheres of interaction to broader issues extending beyond the man-machine interface to encompass managerial and political issues in the organization of complex enterprises. This seminar, in the fall of 1973, established the guidelines for inventing variables that could describe and measure technological and social organizational behaviors in a single frame of reference. That winter, we decided to test the capacity of these variables to describe a complex technological activity by analyzing the aircraft guidance and control system at the Portland International Airport. We are most grateful to the Air Traffic Control Division of the Federal Aviation Administration for permitting us to observe and analyze a complex communication system in action. The results of this field exercise were so encouraging that we continued to experiment with enlarging the scope of analysis of the model.

Dr. John W. Sutherland, of Rutgers University, a Visiting Professor of System Science in the System Science Program of Portland State University, invited us in the spring of 1974 to present our model to a Departmental Colloquium and this spurred us on to tighten the conceptual framework. Soon after this, we were able to complete the first operational version of the full-scale model for publication (Fried, Molnar, 1975). Professor Harold Linstone, head of the System Science Program at Portland State University, greatly encouraged our exploration of "systems" approaches. This accounts for certain cybernetic features found in the model.

At this point we became aware that we were developing a general model that had important implications for planning and policy making for situations where complex enterprises intermeshed technological with social and political factors. The introduction of complex European technologies in Asia provided us with contrasting cases of a Japanese success and a Chinese failure to incorporate these technologies, and, thereby, the model was applied at a policy level of analysis concerning whole societies.

The book emerged, finally, as a result of bringing together all the theoretical issues and results of various pilot studies. It is largely due to the persistent promptings and encouragements of Dr. John Sutherland that we went about preparing this manuscript for publication. We hope it will prove to be a useful analytic tool for discovering dysfunctions in complex systems and in evaluating the merits and failings of alternative strategies of organizing responses to change in these systems.

Portland, Oregon 1978

part I

The Model

1 Introduction: the Problem

The purpose of this work is to show how it is possible for the social sciences to recast its theoretical underpinnings in a manner parallel and analogous to the evolution of scientific theory in the so-called hard sciences, *i.e.*, chemistry, physics, or astronomy. From long and bitter experience it is well known that this task cannot be accomplished by the naive borrowing of the specialized conceptual frameworks and methodologies of these rigorous and highly disciplined sciences. We fully recognize that such theoretical and methodologically sophisticated approaches were developed to consider phenomena of the external world obviously different from the apparently diffuse, variable, and mysterious phenomena of the social and cultural milieu. We agree with Boulding (1956) that we cannot extrapolate or use analogies derived from the physical, or even biological sciences, with their tighter ranges of interaction processes and outcomes, because events in the social system throughout its range fall into a probabilistic pattern and are highly variable. In addition, they have the "Heisenberg principle" built into them in that the strong possibility exists that the process of observation or research procedures can interfere with the phenomena being studied. However, we must state at the outset that we join with those who reject one implication of such a dismaying observation—namely that, this being the case, rigorous scientific study of sociocultural phenomena is not possible and, therefore, only intuitive, aesthetic, or merely descriptive (as opposed to explanatory) theory can apply to what are essentially humanistic topics (. . . that the study of cultural phenomena is to be considered as a historical rather than a scientific-materialistic enterprise: cf. Kroeber's (1952) patterns of historical growth versus White's (1949) technological orientation to explaining cultural complexity).

The supposed intractability of sociocultural or historical phenomena to rigorous scientific analysis, we believe, stems from an inadequate and, hence, inappropriate understanding of the evolution and development of scientific thinking in the Western world. As is generally known, the history of every classic discipline shows that theories of explanation concerning different sectors of phenomena in the physical world have moved from a descriptive and classificatory mode to a relationally analytic and abstract (rather than concrete) approach. The vocabulary of conceptual terms mirrors this change in fundamental frame of reference when qualitative statements become quantitative statements. Accompanying this is a movement away from conceptualizing phenomena as unique events involving complex entities, to simple, abstract, and universally general principles exhibiting basic relationships expressed as rules or laws of behavior (without troubling to define the nature of the entities themselves).

Despite the well-nigh universal acknowledgment of this characteristic of the history of science by contemporary social scientists and the existence of numerous attempts to develop a "science of society" (from Compte to Parsons), the basic shift from substantive, complex, and qualitative conceptual categories to an abstract, relational, and quantitative mode has not been successfully accomplished by those who attempted this task.

The roots of such a failure to accomplish this scientific task are embodied in a series of devastatingly complex and elusive problems recognized by all such theoreticians. The entities whose phenomena are observed, *i.e.,* cultures, social systems, personality structures, etc., are exceedingly complex. They appear to require treatment as complex wholes, which cannot be reduced to simpler component parts for analysis. A "cultural" or "social" system treated as an entity *sui generis* appears, similarly, to contain a series of almost equally complex subsystems, which themselves demand holistic treatment: personality systems, social systems, technological systems, symbol systems, etc. A characteristic of such theoretical exercises is that they exhibit circular reasoning. A complex entity, such as a total social system, is posited which contains as its constituent components conceptually equally complex subsystems such as linguistic, kinship, technological, or other "institutions" each of which is dependent upon the posited existence of the others.

Our point is not to argue the existence or nonexistence of such conceptual entities as culture, society, institutions, symbol systems, etc., but to state firmly that analysis cannot and should not proceed from consideration of the qualities and characteristics of such diffuse substantive categories. We relegate the consideration of the histories and behaviors of such macro-entities to the very legitimate enterprise of humanistic scholarship. Scientific inquiry must proceed by bypassing consideration of the nature of such complex entities themselves, and concern itself with the principles or rules that explain how behaviors emitted by empirically observable entities operate. These, we submit, in all cases of sophisticated explanation in science concern abstract principles of organization.

In seventeenth century astronomy, at last, it was no longer the actual sun, moon, planets, or fixed stars themselves and their apparent movements that became the focus of analysis, but the abstract forces that ordered their movements that received mathematical treatment. In eighteenth century chemistry, a similar change occurred from treating substantive and qualitative characteristics of physical entities over to abstract, relational thinking such as is embodied in the kinetic theory of matter or explaining phenomena as due to the combination of molecules in different configurations.

With these scientific achievements in mind, numerous social scientists tried to achieve a similar transformation of pre-scientific to a fully scientific orientation concerning sociocultural phenomena. The need to find abstract, relational principles of organization of society and culture was always there, yet the theoretical means to accomplish this aim eluded them (despite numerous premature announcements of having achieved the necessary intellectual breakthrough all through the nineteenth century: cf. Morgan, 1877; Marx, 1867; Spencer, 1876, etc.). Their approaches, when examined critically, as we will do in some detail below, all show that to accomplish this "scientific task" they only developed pseudo-relational rather than truly abstract and quantifiable conceptual tools of inquiry and analysis. Those theoreticians who attempted to discover quantifiable measures of sociocultural phenomena tended to utilize concrete attributes of unexplained entities, *e.g.*, White's concept of "energy" and its relationship to societal evolution. Others create empty-sets of pseudo-variables that can have no measurable expression, though quantification is clearly stated to be a needed attribute, for example, Parson's "universal pattern variables". Still others assert fundamental operational concepts which defy the creation of variables to test them, such as Benedict's "patterns of culture", Weber's concept of "legitimacy", or Durkheim's notion of "mechanical" versus "organic" solidarity as distinguishing social structures.

All of these concepts are really classificatory in nature and simply sort and resort unexplained phenomena in different groupings which in fact substitutes description for the analysis of *process.*

Finally, there is that class of pseudo-relational approaches which avoids theory completely at the inception of research and attempts inductively to derive theories about sociocultural behavior from statistical distributions of correlations between traits treated as variables. Most contemporary cross-cultural research efforts tend to fall into this category: Textor, 1967; Spiro, 1965; Murdoch, 1967, etc. Traits of any sort growing out of a descriptive, substantive approach to data cannot perform the function of true abstract, relational variables, which alone are appropriate to the analysis of process. (We do not challenge the utility of treating traits in a statistical manner for some classes of research where a substantive approach *is* called for.) Such descriptive or concrete traits have no bearing on the principles underlying the organization of observed behavior and their manipulation cannot help formulate theories with abstract, organizational properties.

This book presents a model which we believe is capable of dealing with the processes of sociocultural change in which all the above criticisms can be met and overcome. It is our contention that progress in the development of the social sciences has been delayed by the inability of its practitioners to accomplish that necessary transition which happened in the history of certain other scientific disciplines—namely, the ability to recast the world of phenomena into the language of abstract, relational, and quantitatively expressible principles of organization. Because we believe this to be the essential core of the problem, we expended our energies on the search for a solution to explaining sociocultural phenomena based on formulation of a class of variables which can describe organizational properties in abstract relational, rather than substantive, terms. Only such variables can permit first-order, concrete observed phenomena to be recast into a first-order level of abstraction needed for analysis. From other sciences we learned the necessity of avoiding the trap of treating the raw phenomenological data themselves as fundamental units of analysis. Theory requires that such event-systems first be translated into the abstract language of organizational variables in order to become data for a scientific model with explanatory (versus descriptive) properties. With such variables as units of analysis, a true process model becomes a possibility.

Those already acquainted with modern "systems science" orientations toward studying complex, heterogeneous and changeable entities, such as technologies, social institutions, or entire cultures, will recognize immediately that the model we will present utilizes such an approach. By stating at the outset that our model is not concerned with the substantive properties of systematically organized entities but with their *organizational modalities,* expressed as abstract, relational conditions, we have made its fundamental linkage to cybernetics, communication theory, and general systems research clear. As Buckley (1967:1) shows, in the 1950s there was growth of interest by social and behavioral scientists in mastering those new theoretical approaches which "mark the transitions from a concern for eternal substances and the dynamics of energy transformation to a focus on *organization* and its dynamics based on the 'triggering' effects of information transmission."

There is general agreement on a wide front by the new breed of "systems" theorists like Wiener, Rappoport, Campbell, Bateson, Deutsch, etc., that these theoretical frameworks provide much better conceptual tools of analysis than the traditional static, structural, or functional models expounded in the social sciences for treating systemic entities that are complex, fluid, utilize symbol systems, exhibit circular interactive processes, manifest degrees of freedom in behavior, are goal seeking, and are capable of self-regulation and self-direction (Buckley, 1967:2).

Since the early 1950s there has been no lack of exhortations for social scientists to abandon static or outmoded theories and look to modern systems approaches for a fresh start. But despite an increasingly rich literature that provides abstract theoretical conceptual resources for the generation of the new theories to explain complex sociocultural phenomena, there is an actual dearth of empirical research by social scientists who use genuine modern systems models that they themselves originated, rather than borrowed in an eclectic manner from one or another classic statement of cybernetic or information theory.

In designing such research efforts the model utilized must indicate explicitly (rather than in general theoretical terms) specific procedures that permit the translation of empirical data classes into units of analysis which are of an organizational and relational (not static) character. Ideally, these units of analysis should be capable of quantitative expression. The model should also show how and according to what rules of transformation organizational changes being observed occurred. Instead, the social science literature that seeks to examine the processes of sociocultural change, even including works produced by those conscious of the need for a modern systems approach, are still adding to the list of "determinants" that must be included in their process models along with the more traditional components such as society, culture, personality, ecology, institutions, etc. Hence, it remains true today that "social science is made up of partial theories and empirical pieces of evidence with unclear relations among them" (Gastil, 1972:402).

In sum, what is lacking are specific social science models with specific instrumentalities and methodologies which embody the new sophisticated principles and theoretical approaches, but which are also specifically designed to study important aspects of social and cultural theory. Research applying the new theoretical resources to concrete manifestations of cultural phenomena not only must begin with a clear and rigorous statement of how systems theory will be utilized in the formulations of initial conceptual frameworks, but must also *use an entirely new body of data categories* that are not tied to older, classical theoretical or data types. Merely using the old data in new interpretive contexts, we assert, will only prevent the new conceptual categories from being effective.

The aim of this book, therefore, is to treat sociocultural phenomena in the new manner of "systems" approaches, specifically for developing a theoretical model of how technology interacts with social institutions. It departs from the traditional mechanical-structural or functional orientations which, essentially, describe how social entities maintain their forms (*morphostasis,* in the newer terminology). It also eschews those pseudo-dynamic formulations of how social entities change, in which complex subsystems, each with its unique substantive attributes, interrelate to produce new forms *(morphogenesis).* It accepts the challenge of treating both morphostatic and morphogenetic processes in a single frame of reference. Certain basic mechanisms borrowed from communications and information theory, such as positive and negative feedback, and cybernetic concepts that describe system states such as "steady-state", "amplification", or "reduction", have been incorporated in the model. Manipulations of data to form diverse organizational patterns in the form of alternative *scenarios,* model real or hypothetical sociocultural conditions. The data categories used are abstract, relational, and quantitatively measurable variables developed expressly for the type of analysis required and not borrowed from other disciplines.

Such a model has freed itself from the need to utilize, substantively or conceptually, the traditional systemic categories of society, culture, personality, or environment as independent or dependent interacting subsystems. It only treats the organizational implications of what some empirical social entity supplies as *inputs* to an overall organizational design. The model only describes and explains behaviors in relation to the characteristics of this organizational design, and, hence, frees itself from the truly unnecessary task of having to describe or explain, prove or disprove, the existence of such subsystems as entities or processes. Nowhere in this model will there be enigmatic flow diagrams showing "black-boxes" labeled society, culture, personality, or technology and linked by arrows. It is possible to express the organizational implications of behaviors of social entities when they interact via a new class of variables that measure organizational capacity. Only these variables appear as content (data) in the model. No systemic entities *per se,* as units, are treated as content. For example, institutions of a social, political, or technological nature do not appear as such, but their organizational procedures are mirrored conceptually in three *zones,* which themselves are not entities but refer to levels of analysis. One zone displays the immediate organizational pattern involved in the actual carrying out of a specific task (our Technological Zone). Two more zones treat the managerial and political aspects (strategies and tactics) of allocating resources of men and materials to the performance of the tasks being analyzed, and thereby analyze the behaviors and functions of various social institutions (our Managerial and Political Zones).

In analysis of the interplay of these *zones,* it becomes possible to determine if the format of carrying out a particular task can be matched by the organizational capacity of those social entities that are organizing such tasks. Thus, there exists the basis for an evaluative procedure which can tell how *viable* a given society is in terms of its organizational capacity to cope with the array of tasks that make up a cultural corpus of activities. Further, the model can analyze a given society as it is currently organized and predict its capacity to incorporate new technologies. It can measure the gap between present organizational capacity and future desired organizational capacity.

2 Historical Review and Critique

In this chapter we seek to demonstrate by a selective historical review of major theories of social and cultural change that much of the theory and methodology embodied in research studies and theoretical speculation have not dealt with the central issue of discovering and explaining the processes of change. As Barth (1967:661) points out, for example, in many studies static comparisons between two time periods mask as process-analyses of change. Inferences as to the causes or mechanisms that produce the differences noted between two time periods which cannot be substantiated by any research procedure available do not provide insights or explanations about processes. The method of observing rank-order changes in a selected set of social or cultural characteristics over time still cannot evade the label of a static analytic mode since this provides no explanation of what is causing the change and there is no way of making accurate predictions about the course of events. In the history of social scientific theory in this century, frustration of some theoreticians over failure to overcome this critical stumbling block of how to explain *process* while treating sociocultural data of an essentially concrete and static nature has led them to seek for dynamic causes of change in social psychological terms. For example, critics of traditional anthropology's preoccupation with changes in the forms of social institutions, such as families, descent rules, value systems, technological systems, have shifted to a *first-cause* argument in which it is *people behaving* that causes sociocultural phenomena to come into being. Those not committed to such a reductionist ontology fall into the so-called eclectic category in which dynamics of culture change is viewed as the mutual interaction of a series of subsystems, generally identified as culture, society, and personality (Linton, 1940; Mead, 1956; Kluckhohn, 1944; Parsons, 1951). All such approaches tend to be tautological in that each subsystem can only be explained in terms of the other. True, the deterministic first-cause explanation is eschewed, but at the cost of failure to provide dynamic explanations of superior intellectual merit.

We believe (and thereby agree with Murdock, 1971) that at the heart of the matter is this ontological confusion arising from the misplaced need to identify and define the nature and properties of such entities as cultures, societies, institutions, basic personality type, etc. The existence or nonexistence of such entities is not the real issue to be confronted in developing a process model for social change. It is not only unnecessary, but it is a mistaken approach to begin with undefined complex entities as the source and cause of the phenomena treated in studies of the processes of cultural change. At this much too high level of abstraction, you cannot develop operational procedures to identify or discover any "agencies" that cause change. In addition, the futile search for "efficient" causes of change is endless because the researcher is swamped with empirical data of easily identified causal agencies in a given concrete historical example: great men, new religions or ideologies, technological innovations, migrations, wars, geological catastrophies, etc. What is lacking, therefore, in both approaches is that theoretical point of departure which focuses upon the manner of discerning the organizational forms that sociocultural phenomena fall into and their transformational properties as embodied in rules governing their behavior as a pattern of relationships. The consideration of specific "causes" operates on the level of treating the substantive elements (generally identified as components of such subsystems as society, culture, personality, ecology, etc.) which must combine in order to produce sociocultural phenomena. Therefore, at higher levels of theoretical explanation, causes and the search for causes must be left behind and real event-systems can only serve to illustrate the operation of general principles of combination such that their behavioral characteristics stem from their organizational properties and potentialities.

In what follows, we will see that earlier influential theoreticians have failed to deal adequately with the mechanisms or processes of change because they never succeeded in freeing themselves conceptually from empty abstractions or statically treated phenomenological events involving undefinable complex substantive entities. Murdoch (1971) has radically rejected such entities as society, culture, or personality types as classic examples of the fallacy of misplaced concreteness to be treated as "myths", as intellectual fabrications of no real scientific utility in grappling with truly fundamental issues of process. Unfortunately, he joins writers like Barth (1967) who look for a better theoretical starting point in the equally futile reductionist position that it is in the realm of real people behaving that the new process models will find their data and dynamic explanations. In contrast, the thrust of our conceptual thinking is toward treating the forms that change takes and how such organizational forms act as *parameters* within which the characteristics of real events can be explained and predicted.

A major problem for social scientific concepts of change has been the issue of levels of analysis. Some theoreticians, like Morgan (1887) and White (1959), the so-called macro-evolutionists, have treated the entire history of mankind and culture within a single frame of reference; hence, their conceptual vocabulary is highly abstract and general. The neo-evolutionists, like Steward (1955), Harris (1968), or Sahlins (1958), appear to lower the level of abstraction and treat the development of specific societies over time and search for uniformities to illuminate causal interpretations of change. Finally, theoreticians like Barnett (1953), Wallace (1956), Barth (1963), Goodenough (1963), or Parsons and Bales (1955), focus attention on the dynamics of interaction processes at the personal and interpersonal level in order to understand the dynamics of culture change. Actually, many anthropological writers are quite willing to recognize the need to interrelate and mutually explain how all three levels might interact since they are not held to be mutually exclusive realms of inquiry (Kluckhohn, 1944; Mead, 1964). Parsons (1960) and Lasswell (1948) are outstanding examples of sociologists undertaking this very task. It is no wonder that such writers are driven to seek in interdisciplinary approaches that which their starting discipline's stock of theory failed to supply (Linton, 1939; Kluckhohn, 1944; Hallowell, 1948; Parsons and Shils, 1959, etc.). At the higher levels of abstraction, traditional anthropological concepts of "culture" or sociological concepts of "society" might serve, but at the more concrete levels where a search for "prime movers" is conducted, psychiatry or social psychology may appear to be methodologically and theoretically better equipped disciplines.

Related to the issue of discovering ontologically causal entities is the long-standing theoretical debate as to the causal *priorities* to be assigned to culture, society, personality, technology, or natural habitat in effecting change. For example, culture viewed as symbol systems, ideologies, or value systems are considered as dynamically "prior" in the writings of Kroeber (1944), Weber (1946), Benedict (1934), Mumford (1954), Toynbee (1947), Sorokin (1937–41), and Turner (1974). The English social anthropologists, on the other hand, like Radcliffe-Brown (1952), Fortes (1945), or Gluckman (1968) (obviously siding with Durkheim in this matter) and sociologists such as Merton (1968), Dahrendorf (1959), and Goldner (1963) display writings which assert the theoretical priority of "society" or social institutions. Again in contrast, writers like Barnett (1953), Wallace (1956), Goodenough (1963), Brown (1966), and Adorno (1950) have stressed the causal importance of psychological factors in inducing change. Finally, the techno-ecological approach as presented in the writings of Steward (1955), Harris (1968), Sahlins (1960), or Service (1971) suggests yet another sector of "priority" as factors initiating and/or conditioning processes of change.

We believe the inability of these writers to resolve the issue as to which sector of analysis has priority stems from their faulty statement of the problem. The sector of behavior or phenomena assigned causal priority (and this criticism holds for the "eclectic" position of seeing mutual interaction of a series of different sectors) nevertheless remains recognizable as referring to substantive entities, the properties or attributes of which require for their definition the prior existence of others which it is supposed to influence or order. A typical example is provided in Kardiner and Linton's (1944) concept of "basic personality type". This typological entity is developed out of a process of socialization which has its forms governed by social institutions which, in turn, demands a unique culture for their maintenance, and the unique culture is a product of the basic personality structure. The circularity of the argument is obvious.

A third concern of culture change theory has been to determine whether or not what is observed as changing is doing so within a cyclical, repetitive pattern or is moving in a nonrepeatable, irreversible path. Leach's (1965) classic analysis of the Highland Burmese tribal political system in which Kachin tribesmen radically shift political structures back and forth between two strongly contrasting types—the hierarchical and the egalitarian—was meant to challenge traditional notions of drastic change as being unidirectional. This represents the research tactic of finding a significant exception to disprove a theory without necessarily offering a new theory. With Spengler (1927), and even to some extent, Kroeber (1944), societies or cultures exhibit developmental stages in a kind of natural history of morphological development, which on the surface appears to be evolutionary change but in fact is only part of a cycle which will come to an end. In contrast, Morgan (1877), Tylor (1865), and Marx (1967) formulate a universal process of social change where there is progressive changes in a well-defined direction and which is not cyclical in type. Attempts to demonstrate in operational terms how we can decide which alternative mode of change is taking place while it is happening, and not in retrospect, have not been successful. The fundamental reason for this failure is that the units of analysis undergoing change, whether society, culture, technology, or ethnic groups, are in all cases treated as though they have substantive properties and a true process model cannot be developed to treat such conceptual entities.

Another problem of culture change theory, perhaps more vexing to the structural-functionalist partisans such as Radcliffe-Brown, Fortes, or Gluckman than others, is *persistence* of apparently dysfunctional or nonfunctional social elements in a society under conditions of change, and how to explain the maintenance of the status quo despite considerable accumulations of change. If your theory forces you to give equal weight to all existing cultural traits and find "functional" attributes for all aspects of an ongoing social process—*i.e.,* making them necessary parts of the total social pattern—then retention of historically older elements in the context of entirely new conditions which appear as anomalies must be explained somehow as *normalities* within the social system. A visit to any American Indian reservation will demonstrate that all sorts of traits and behaviors can coexist without being part of a single functionally viable society. Apparently, almost any bizarre or exotic set of behavior, even those generating extreme community conflict, can be maintained in the face of grotesquely inadequate social integration if what is missing is supplied from "external" sources such as the specialized agencies of the national government. Crudely stated, the principle implied as a challenge to classic structural-functional theory is this: If people eat and reproduce, they can have very sloppy sociocultural characteristics. (Or the trite truism that some societies are better organized to cope with their problems than others!)

The work of Barnett (1952), Wallace (1956), and Goodenough (1963) dealing with the psychological reactions of people in societies undergoing acculturative change, while making a contribution in emphasizing new situational aspects to be considered, is of limited value to the major issue of defining where and what to look for as explanations of the processes of change. Messianic or revivalistic cults are reactions to some kinds of change wherein members of a society may be struggling to re-establish some measure of stability in a rapidly deteriorating sociopolitical environment. But such reactions to change cannot tell us what brought about the change or even why not all apparently badly organized social situations are perceived as intolerable. Statements of the problems involved in rapid culture change in terms of social psychological parameters may or may not be legitimate terms of references for some anthropologists, but we feel that a much more fruitful approach is to discover the principles that determine what are viable or nonviable organizational conditions for the survival of a given society in its existing form when the need to change becomes imperative. We still do not know why different societies manifest different capacities to tolerate change and resulting inconsistencies that emerge in the fabric of its social or ideological structures.

The above discussion represents what we feel to be an exposition of the fundamental issues that must be considered in order to understand why an adequate theory of the processes underlying sociocultural change has so far not emerged. The fundamental characteristic of all essentially substantive and classificatory orientations to theories of change is their positing of the real existence of systemic entities such as cultures, social structures of personality types. We have identified four problem areas that emerged for theory in considering phenomena of culture change that stem from the positing of pseudo-substantive entities: (1) the levels of analysis problem; (2) the ontological assignment of causal priorities among the interacting "systemic entities"; (3) the problem of identifying change as cyclical in character and occurring within a pattern versus progressive, nonrepetitive change into entirely novel patterns; (4) the problem of persistence of elements within the context of change which challenges the concept of necessary functional integration of aspects of a social system.

The purpose of this exposition of social scientific theory has been to heighten awareness of a single central point. The strength of social science has been in exposing and elaborating various sources or causes of change of an empirical nature. The weakness has been in failing to develop a theory that shows how and by what rules or principles such sources or causes move the organization of behavior in a given society from one pattern of relationship to another. We heartily agree that the investigation of specific causes of change falls into an indeterministic, and hence humanistic, category of investigation. Whether or not great men arise, culture contacts occur, technological innovations be a response to the challenge of nature's whims, etc., are indeed matters of great moment for a history of specific societies. But regardless of the source, identity, and nature of such inducers of change, in order for them to have an impact on sociocultural institutions, they must be translated into programs of action that have organizational characteristics. It is the investigation of these organizational characteristics that is the fundamental starting point for our venture in theory and method.

Since many brilliant and classic theoretical investigations by such scholars as Durkheim, Weber, Marx, Redfield, Steward, Parsons, etc., have indeed also started from such a fundamental premise, it is necessary to state that we will suggest a novel solution to the problem of devising a true process model that does not borrow its methodologies or theoretical assumptions from any of these other theoreticians.

3 A New Process Model for Technology and Sociocultural Change

Our critical review of culture change theory has indicated that theoretical formulations concerning the processes of culture change that adequately describe and predict the course of events under conditions of change have not emerged. Basically, theoretical failings fall into two categories: formulations that are too abstract and general to have empirical content or be subjected to tests of verification; and those that indeed are rich in empirical contents but fail to produce generalizations about underlying processes of culture change.

Therefore, our model seeks to overcome these twin difficulties and achieve a solution that both is rich in empirical content and does generate principles or rules about conditions of change that are universal in application and are verifiable by explicit, operationally defined procedures.

To accomplish this end, it was necessary to abandon specific theories and hypotheses stemming from any of the traditional social scientific disciplines such as anthropology, sociology, or political science and, at the same time, develop a new class of generalizations equally applicable to the subject matters of these various disciplines. Hence, our approach is neither unidisciplinary nor multidisciplinary. Since each of the traditional disciplines looks at classes of phenomena in specialized and unique ways, their specific theories and methodologies were explicitly ruled out in our search for a novel overall frame of reference which would permit new theories and hypotheses to emerge.

We believe such an enterprise is theoretically possible because there is a common ground underlying the apparently diverse and substantively unique subject matters of these disciplines. Once we shift emphasis from the concrete or substantive subject matters themselves, *e.g.*, their attributes, qualities, or characteristics, over to a purely relational mode of considering social phenomena, the common feature that emerges is that of *organization itself.* Organization by itself would constitute yet another empty abstraction, a futile truism, if it were not for the fact that all forms of organized activities in any sector of sociocultural phenomena exhibit patterns of *constraint.* The *measurable* characteristic of organization is the constraint that emerges from the configuration of the participant elements of that activity which is organized. The specific content of these elements will vary when the activities being treated are of an economic, political, social, or ideological nature, but not the principles of organization—these are unaffected by the empirical character of the constituent elements.

It is the principles or rules of organization that, ultimately, govern the behavior of technological, political, ideological, and other sectors of activity. Thus, these principles or rules of organization constitute the basis of a nonsubstantively conceived, general social science process model.

In developing our model we sought to meet the following criteria of a true process model:

1. It must have the power to describe, explain and predict the operations and course of events of activities as they occur over time.
2. It is not activity-specific. It can treat organized behaviors of different contents in the same frame of reference.
3. It measures and interprets the significance of changes in organizational format throughout the field of interacting components that make up an activity (that is, produce a holistic analysis).

4. It permits the cross-cultural comparison of cultural phenomena by converting all empirically observable elements into a single, contextually homogeneous frame of reference so that the specific identity of classes of cultural elements is no longer relevant to analysis.

To achieve the goal of creating a general model with such powers, two fundamental conceptual and procedural breakthroughs had to be accomplished. First, the traditional and substantively conceived categories of society, culture, and technology had to be abandoned. In their place new categories whose contents display the dynamic and relational properties required for the model had to be developed. After a long process of trial and error involving both inductive and deductive procedures, we came to believe that the traditional categories of technology, society, and culture could be referred, empirically and operationally, to three *zones of organizational activity.* Our master concept of "organization" could not remain an empty abstraction but in the model must be operationally referred to some empirical activity which is the object of analysis. It does not and cannot be made to treat second-order abstractions such as society, culture, or personality directly.

These "zones", therefore, were developed not to represent different categories of activities of an economic, political, religious, military, etc., variety, but refer to a class of relationships involved as fundamental aspects of organizing activities. The treating of some specific task or tasks is, then, the core of the model. Rather than developing a new vocabulary, we chose three well-worn but serviceable terms to stand for organizational involvements at three levels. The zone of "technology" represents the immediate interface of men and/or artifacts in some combination, organized in some configuration in order to carry out some concrete task. Next, instead of using the classical terminology of the social sciences to identify specific social structures as family, kin, tribe, corporation, state, political party, etc., to refer to the way in which a social field of persons (social organization) interact according to rules (ideology or value systems) which assign the right persons and resources to accomplish a task (institution), we found it more useful to refer to a zone of organizational activity centered on the *direct management* of an activity (modelled in the Technological Zone)—hence, a *Managerial Zone.*

Another more complex level of organization occurs when the tasks requiring "management" are so complex and linked that levels of management occur; we modelled the organizational activities of this level in a *Political Zone.* In this manner, we hoped once and for all to eject the substantively conceived systemic pseudo-entities of the traditional vocabulary of terms from the model. Only the specific empirical activities themselves would be modelled and not economic systems, political systems, social systems, etc.

If it should be questioned at this point whether this substitution of three zones for the traditional systemic conceptual sociocultural entities is merely a substitution of one pseudo-relational set of terms for another, our answer is that the zones of activity, unlike the other traditional categories, can be operationally defined and described by a class of variables which can measure organizational characteristics. Hypotheses and theories associated with the application of the model can be tested and verified by explicit procedures set forth in the model.

Therefore, the second requirement of the model is that it utilize authentic variables that *measure* organizational features of empirical activities (termed "tasks" in our model) that make up zones. We will begin the exposition of the model with a presentation of this new class of variables.

The Variables of the Model

Two features of these variables are required if they are to meet the specifications set by a true process model: (1) the ability to measure any and all empirical conditions treated by the various social science disciplines; (2) the specification of a set of operations whereby the interaction of these variables can abstractly describe the pattern of relationships which characterizes any technological or sociocultural activity. The ways in which these relational patterns change in response to changes in the loadings of the variables must follow from such sets of operations.

The variables utilized in the model were developed through a long inductive research effort. To ensure that all technological activities, ranging from the simple to the complex, would be measured equally well by the same variables, an extensive survey of modern industrial, communication, and information processing systems was carried out in the Portland, Oregon area. (A limited aspect of this exploratory research activity is reported in Fried, Weitman and Davis, 1972.) The first complete set of variables was pretested in the study of two complex man-machine systems at the Portland Airport. (The analysis of the Air Traffic Control system is used in chapter 4 as illustrative materials.)

A similar approach was taken to developing the variables to describe and measure the Managerial and Political Zones. The social institutions of a range of societies, simple to complex, in different culture areas of the world were evaluated and classified in terms of their organizational attributes as associated with the specific techno-economic tasks they were carrying out. Only when a variable could apply to any format or organized response in any and all of the societies or cultures examined was it considered ready for testing in the model.

For organizational analysis, the task categories selected were very basic techno-economic activities found in all societies, for example, production and distribution of vital goods and services, transportation and communication. Only when a generalized organizational variable was found to be adequate to describe and measure specific managerial or political behaviors with reference to these tasks and for a range of societies, simple to complex, traditional or modern, was it considered ready for testing. The way in which empirical, technological, managerial, or political behaviors are operationally translated into organizational variables will become apparent when the definitions and descriptors of the variables are given below.

What follows is an exposition of eighteen variables, nine of which describe and measure technological conditions, and nine, social organizational conditions. It is hoped that this enormous compression will permit the vast ranges of empirical phenomena encompassed by these two aspects of culture to be disciplined into simple and easily visualized categories. The variables developed measure degrees of constraint along a continuum. Changes in their values are meant to reflect changes in real behavioral conditions of empirical entities since the variables are assigned loadings, low to high, with reference to observed behaviors of men and artifacts in various combinations performing technological or other culturally determined activities requiring organization. The method by which individual variables are combined in order to form a picture of the organizational pattern of a technological or social activity is given as a set of explicit operational procedures. Following this, some hypotheses by which any desired new organizational format may be achieved by changing the loading of specific variables (that measure the component parts of the activity being analyzed) are suggested.

Variables as Measures of Constraint

The relational approach to the analysis of organizational patterns has as its basis the concept of constraint. Constraint in our model refers to the degrees of freedom that are mutually imposed by the variables as they interact in forming organizational states. Thus in an organizational format where there is high constraint, there is a corresponding low degree of freedom, that is, the possible actions that can occur within such an organizational state are limited and generally highly structured. Where there is low constraint, the resultant organizational state is relatively open to a variety of activities within its parameters.

There is no normative aspect implied in interpreting modes of organization as being more or less constrained and thereby superior or inferior. Only a descriptive measurement of a given organizational format

Table I. The Variables

- I. TECHNOLOGY
  - A. The Man-Artifact Component (M/A)
    1. Locus of Dominance Variable
    2. Complexity of Assemblage Variable
    3. Phases of Unit Interaction Variable
  - B. The Task Component (T)
    1. Serial Characteristic Variable
    2. Operations-Output Relations Variable
    3. Output Form Variable
  - C. The Setting Component (S)
    1. Setting Structure Variable
    2. Locus of Input Variable
    3. Autonomy Variable
- II. SOCIAL ORGANIZATION
  - A. The Communications Component (C)
    1. Locus of Authority Variable
    2. Channels of Communications Variable
    3. Complexity of Linkage Variable
  - B. The Domain Component (D)
    1. Size-Quantity Variable
    2. Heterogeneity Variable
    3. Concentration Variable
  - C. The Legitimacy Component (L)
    1. Boundary of Authority Variable
    2. Social Control Variable
    3. Locus of Validation Variable

in terms of constraint characteristics is involved. As we will see later, whether or not an organizational format requires more or less constraint for an effective interface with a given task depends on the characteristics of that task.

The variables by themselves cannot be used in analytic exercises until they are grouped into *components.* The component is the minimal unit for describing a pattern of constraint. Thus, a triad of variables, each measuring some specialized aspect of constraint associated with the behavior of an empirical entity, is analyzed as to its format of organization in a component. (A minimum of three points is required to establish a pattern.) The various activities of technology and management are, in turn, modelled by the interaction of these components. There are three such analytic components used in the model for each sector of analysis (*i.e.,* zones). The use of three units is required for two reasons. Firstly, triads form the simplest identifiable pattern. Secondly, our conception of an "organizational state" relevant to treating the materials of technology and social organization requires that we model the interactions of three aspects of any activity: (a) an acting entity (or entities), (b) the action being performed, and (c) a setting wherein the action occurs. Thus, while the variables were developed inductively from an examination of technological and social institutional behaviors, the analytic units, the components, were more deductively derived from our conception of an action process involving the above three elements.

In order to treat the interaction of technology and social organization, we now require the term *zone.* A zone is a relational mode composed of components that describe one of three types of organizational activity. These three zones are: a technological zone, a managerial zone, and a political zone. The technological zone describes those organizational states that relate to the manipulation of the physical universe, usually through the means of artifacts (tools, machines, etc.). The managerial zone involves the organization of those activities that make up a given task. The political zone concerns the articulation of numerous managerial activities into a coherent and viable system.

The technological zone is the area that tends to exhibit the highest degree of constraint. The constraints here stem from the fact that the physical universe and the disciplines imposed by the use of artifacts limit the freedom of responses possible when men organize these activities. The managerial zone encompasses activities which no longer have a direct interface with the physical universe or artifacts, but organize and make viable the flow of activities in technological behavior and thereby primarily concern the human, interpersonal dimension of interaction.

The political zone involves, literally, organization of organization. It orders the managerial activities that more directly interface with technological activities. It is thus twice removed from the constraints that are imposed on an operational level involving men and their tools and instruments.

The above paragraphs have served to indicate how the variables used in the model are grouped. In the discussion of the variables that follows, they are presented as continua. The range, from high to low value loadings possible within these continua, reflects high and low conditions of constraint. Thus a high loading reflects high constraint for that situation that the variable describes. The value assigned to a variable is made with reference to the empirical situation being described and is not derived from some absolute scale.

Finally, we are indeed aware that there are many possible relational variables other than the ones we developed. However, we hope to demonstrate in the case studies presented in Part II that the variables we utilize are capable of modelling the organizational properties of very complex systems.

Variables of the Technological Zone

THE MAN-ARTIFACT COMPONENT (M/A)

In considering the utilization of artifacts in accomplishing a task in relational terms, it is necessary to treat the artifact and the man or men using it as a single, indivisible entity. Our concern is only with the resultant behavior emitted in their combined actions. These actions are reflected in the three variables that constitute this component.

Locus of Dominance Variable. This variable measures the control and manipulative aspects of the man-artifact unit in the performance of an activity. Low values (*i.e.,* low constraint) refers to a situation where there is a relatively open set of possible actions for the unit to manifest (*e.g.,* a man using a simple multipurpose tool like an ax). High values (*i.e.,* high constraints) refer to situations where the realm of possible actions is limited to a particular one (*e.g.,* the utilization of a program for automated production which rigidly constrains the operative elements to a set pattern).

Complexity of Assemblage Variable. This variable measures the quantitative aspect of man-artifact units involved in an activity. A critical feature of applying this variable is to make the unit of analysis isomorphic to the task definition. Some tasks are made up of subtasks. Either a subtask (*e.g.,* a single man operating a machine such as a lathe) or a total task's parameters (*e.g.,* an entire machine shop considered as a unit) can be treated in the model so long as the level of analytic unit is maintained throughout all analytic operations. Low value assignments refer to a single man-artifact unit. High value assignments refer to assemblages of man-artifact units that interact in the context of a task.

Phases of Unit Interaction Variable. This variable measures the sequential aspects of the manner whereby the man-artifact units are linked in the activity (*e.g.,* in the manner of a PERT analysis). Low values refer to simultaneous actions which describe an immediate and total result that completes a task. High values measure multiphased activities wherein a series of man-artifact elements interact sequentially over time to produce a result.

THE TASK COMPONENT (T)

Technological activities take place within the boundaries defined by the task's specifications. Traditionally, the subject matter of the various social sciences has been determined by a particular class of tasks with each class exhibiting unique properties (*e.g.,* political, economic, etc.). Our concern is with the form that tasks attain as the tasks are mirrored in organizational states, irrespective of their substantive properties. A task's properties in our model derive from its organizational structure, and not from its classificatory status. The Task Component is described by the three variables listed below.

Serial Characteristic Variable. This variable measures the temporal aspects of the mode of attaining task completion. It should not be confused with the Phase variable of the Man-Artifact Component; there, sequential behavior is of concern. While all sequences have temporal dimensions, all temporal activity does not exhibit sequential steps in its performance. Low values refer to situations where the task completion follows immediately upon a man-artifact action (*e.g.,* as in turning on a light fixture with a switch). High values refer to situations where task completion requires a set of actions over time (*e.g.,* completion of the construction of a building).

Operations-Output Relations Variable. This variable measures the degree to which operations that produce outputs—and the outputs themselves—are characterized by their identity or by their separation. By outputs is meant the actual products that result from an observed technological activity. Low values denote situations where an immediately sought output is effectively coterminous with the operations that produce it (*e.g.,* a sail propelled by the wind). High values refer to situations where outputs are separate in time and space from the source of their production (*e.g.,* a missile system, where the hitting of a target is a delayed and extended output of a complex operation).

Output Form Variable. This variable measures the degree to which task operations are capable of routinization. Output, in this context, again refers to that which is produced by an observed technological activity. Low values refer to those procedures that are not subject to a routine and are thus improvisational in character (*e.g.,* devising an experimental model). High values refer to situations in which task procedures are always repeated in a routine fashion (*e.g.,* mass production routines).

THE SETTING COMPONENT (S)

It is important not to confuse the Setting Component with a traditional concept of ecology. Since the Man-Artifact Component concerns the implementation of action and the Task Component defines the act itself, the Setting Component defines the conditions wherein action occurs. The Setting Component is described by the three variables explained below.

Setting Structure Variable. This variable measures the means by which inputs are allowed into the arena of action. Inputs refer to all the various resources that must be combined to initiate a technological activity, that is, the requisite resources. Low values measure indiscriminant or uncontrolled inputs, that is, random inputs (*e.g.,* the number of unscheduled aircraft entering the airspace of an airport). High values refer to highly structured situations where the inputs are tightly controlled (*e.g.,* an automated oil refinery).

Locus of Input Variable. This variable measures the spatial distribution of inputs (*i.e.,* the requisite resources). Low values refer to situations where all of the inputs are highly localized in origin (*e.g.,* a cottage industry such as cheesemaking). High values refer to situations where inputs come from diversified places and so require more constraints in the system collecting them (*e.g.,* the steel industry).

Autonomy Variable. This variable measures the autonomy of the unit being analyzed; it determines to what degree a given task-defined unit contains within itself the inputs necessary for task completion. Low values indicate a low degree of dependence on sources of inputs outside of the unit, that is, self-sufficiency (*e.g.,* a bow and arrow as a weapon). High values indicate a high degree of dependence on multiple sources of inputs and, hence, low self-sufficiency (*e.g.,* an aircraft carrier and its attendant logistical support system).

Variables of the Two Zones of Social Organization

We consider both the Managerial and Political Zones to be aspects that model social organization and are amenable to delineation by the same components and variables. There were two tasks to be accomplished in developing the variables of these zones. Firstly, the variables would have to be of the same order and type as those developed for the Technological Zone because: (a) all components of the model must fit into a single frame of reference, and (b) they must reflect the fundamental assumption that technology is an aspect of culture *per se* and is not a distinct or autonomous sector only attached to social or cultural entities. Secondly, the application of such variables to empirical situations must avoid the need to specify and describe the special characteristics of such distinct social entities as kinship, political or religious institutions, or of the ecological, technological, or ideological forces that condition the resources available to such social entities—and yet they would be able to measure the results of the interaction of such "forces" as a pattern of organization.

Thereby, for example, an examination of the variables of these two zones will show that the Communications Component can express organizationally the actional properties or potentialities of social institutions such as a clan or a political party equally adequately. The Domain Component contains variables that express the actional potentials of various classes of techno-economic institutions that produce the resource base characteristic of different economies (*e.g.,* agricultural versus industrial types). The Legitimacy Component clearly will be seen to reflect that realm of concerns called ideology, symbol systems, or social control. Thus, just as with the development of the variables of the Technological Zone where the full range of transportation, communication, and productive activities were considered, the rich data of the social sciences were not ignored. The variables would surely fail to provide an adequate analytic inventory of tools if these could only reflect a narrow range of inputs that make up the total spectrum of significant interacting elements that produce sociocultural events. Whether or not the small number of variables we developed so far will serve to construct hypotheses and theories that make genuine contributions to the study of social change can be judged from the content of the chapters that follow.

THE COMMUNICATIONS COMPONENT (C)

This component groups the variables concerning man-man interaction relationships. Our assignment of man-man interactions to a communications function is based on two assumptions. First, human actions, even when not constrained by technology, exhibit order. Second, the nature of human interaction modes is profoundly characterized by communicative processes. For communications to occur, information rather than noise must be transmitted. It follows that communications processes and human behavior, insofar as they exhibit order, are organizational in nature, and so are amenable to display as system states in our model. We describe the Communications Component by three variables.

Locus of Authority Variable. Analogous to the Locus of Dominance variable in Technology, this variable measures the control and manipulative aspect of man-man interactions. The unit is either an individual or a group treated as a single unit, since only the emitted behaviors of the total unit are of concern. Low values measure situations where authority is shared or diffuse (*e.g.,* the males in a hunting band). High values measure authority that is concentrated within a single entity (*e.g.,* a dictator or a totalitarian political party).

Channels of Communications Variable. This variable reflects and measures the actions of the number of units being organized and the effectiveness of that organizational structure. The number of communications channels is in positive relationship to the number of units that have to be integrated to ensure the performance of a task. Low values express the behavior of an entity as a single autonomous unit exhibiting a low degree of integration within it, if it is a group (*e.g.,* the autonomous households of a band). High values measure the ordering of assemblages of units that show a high degree of integration and interdependence in carrying out a task (*e.g.,* the administrative structure of a nation-state).

Complexity of Linkage Variable. Analogous to the Phase variable in the Man-Artifact Component, this variable concerns the sequential aspect of management. Low values refer to interactions that are accomplished in a single phase at one time (*e.g.,* ordering a group to cross a river in a boat). High values are assigned to activities that are multiphased and require sequential linkage over time and space (*e.g.,* a bureaucratic decision to implement a five-year plan).

THE DOMAIN COMPONENT (D)

Societies exist in time and space and are subject to a variety of situations that require varying responses. This component measures the adaptive capacity of a given social organization to meet these contingencies. There are limits to any organization's capacity to expand or contract its internal arrangements in response to events. The variables of this component reflect this capacity.

Size-Quantity Variable. This variable measures the dimensional scale or quantity of elements that the organization seeks to control or manipulate. Low values refer to individual or small units (*e.g.,* a small independent grocery); high values refer to the need to organize numerous and/or large units (*e.g.,* a large nationwide chain store).

Heterogeneity Variable. This variable reflects the qualitative aspect of an organization's capacity for acting by measuring the degree of heterogeneity present in the units being organized. Greater heterogeneity of the units in an organization creates a greater potential capacity to respond to a variety of situations since units tend to be task-specific in their behaviors. Thus, if an organization has a heterogeneous array of units in its composition, the organization has a greater capacity to meet and deal with a greater variety of possibilities. Heterogeneous organizations thus require greater constraint than do homogeneous unit organizations, and the latter are assigned low values (*e.g.,* an infantry squad of riflemen). High values are assigned to organizations displaying heterogeneous structures (*e.g.,* an aircraft bomber crew).

Concentration Variable. This variable refers to the spatial aspects of managerial activity; it measures the capacity of the managerial entity to organize units distributed in space. Low values are assigned to activities that occur in narrow, localized settings (*e.g.,* a parish priest). High values refer to activities that must occur as distributed in space (*e.g.,* the Joint Chiefs of Staff).

THE LEGITIMACY COMPONENT (L)

This component represents the ideational analogue to the Setting Component in the Technological Zone. It defines the axiological boundaries wherein the organization of activities occurs. As such it delimits the area of effective control. We measure these boundaries with the variables described below.

Boundary of Authority Variable. This variable indicates the number of areas of jurisdiction encompassed by the managing entity. By area of jurisdiction we refer to those activities traditionally termed institutions (*e.g.*, religious, political, etc.). Low values refer to a single area of jurisdiction (*e.g.*, a teacher in a classroom); high values reflect greater and different areas of jurisdiction being included within the managerial sphere of control (*e.g.*, the medieval papacy).

Social Control Variable. This variable measures the amount of coercion (physical or psychological) utilized to ensure the compliance of managed units. Low values refer to low coercive potential (*e.g.*, as exercised by respected persons from whom one seeks advice voluntarily). High values measure high coercive potential (*e.g.*, a secret police organization).

Locus of Validation Variable. This variable refers to the validity of the control being exercised. Its measurement refers to the source of this validity as expressed in ideological or rhetorical terms for and by the entity being organized. Low values measure personal power not based on an abstract outside principle (*e.g.*, a war chief). This is a low constraint situation since a single individual or small group basing its validity on personal power can only exercise that valid authority within a confined or temporary situation or a person's lifetime. High values, on the other hand, are assigned to conditions where the validity of a claim to legitimacy is expressed in terms of an abstract principle (*e.g.*, a Supreme Court judge).

Modelling System States

As we have noted in the previous section, the variables are continua and so express different patterns of organization when they have different loadings. For the sake of expositional economy, we have been forced to use the two polar values of *high* and *low* as measures of the variables instead of a multiunit scale. The model when it uses only two loadings for a variable is far simpler to use and comprehend than one in which the variables are more sensitively scaled. Since we did not have computer facilities, we were forced to use simplified categories, few in number and of strongly contrasting types. With only two possible value loadings for each variable, each component of a zone will have only eight possible states of organization among the variables. Similarly, the three components interacting as triads will only reflect eight organizational states for a zone. In this form, then, the tables and charts used in the illustrative examples are easy to inspect and operations of the model are more easily followed.

Thus, for broad descriptive purposes and to illustrate the dynamics of the model, this simple scaling device is sufficient. But for prescriptive purposes when, for example, the model is used to diagnose the specific sources of difficulties in a system or to evaluate the viability of a set of alternative organizational conditions, a more complex set of scales to measure variables must be provided. We have, in fact, already devised a ten-point scale for each of our variables. However, the application of these more elaborately scaled variables to specific research projects is only now underway and results will be reported when these projects are completed.

Obviously, these new sets of scales geometrically increase the range of possible combinations for the production of organizational formats. Only a computerized program for generating a full array of possible organizational states can deal with this problem. However, it is not always necessary to generate, via a computer program, all the permutations and combinations of variable loadings for *all* possible future conditions. If planners can specify a limited range of scenarios of future conditions to be analyzed and can supply descriptive information of the probable technological, managerial, and political resources, then these selected scenarios can be analyzed and compared and contrasted with current conditions utilizing manually applied procedures, even when variables are measured by a ten-point scale.

Nevertheless, despite the above disclaimer, we must confront a potential serious challenge to the general utility of our variables for future cross-cultural comparative research. For the limited instance of treatment of simple dichotomous contrasts in degrees of organizational complexity such as pertained in our chosen research illustrations, the issue of validity of the scaling given to the variables has been crudely handled by the technique of having two or more independent observers code the same behaviors. This clearly will not serve as a validating procedure where discriminations could only be useful when related to a multiple-point scale.

The future development of such refined scaling procedures for variables brings in its train two severe problems. One is a cross-cultural aspect in which the uniformity of field, or context, of observation is empirically different from case to case (Japan versus England). A second concerns the vast multiplication of system states that the new combinations create as ideal types. Our exposition of the principles of the model, fortunately, does not require the utilization of such refined scaling procedures, or the entertainment of more than a simplified, eight system state scheme for modelling the technosocial changes occurring in the chosen case studies.

In applying the model to actual case studies, the component level is the first meaningful interactive analytic level of analysis. Since components are the basic units of zones, which represent the observed action systems, it follows that our procedure requires that empirical measurements, while made at the variable level, must undergo analytic transformation into components. The three components describe a zone of operations. The patterning of relationships among the components of a zone thus provides a synoptic statement of the organizational state for either technological or managerial activity. A comparison of the zones permits the assessment of how well the managerial and political organizations interface with that of technology and among themselves, and the specification of which component and specific variables require manipulation to achieve different particular organizational patterns.

We shall illustrate the application of these procedures for treating the three variables of a single component, and then for analyzing a zone with its three constituent components. Table II demonstrates the proce-

Table II. Organizational States Possible for a Single Component of the Technological Zone

	Analytical Pattern (value loadings assigned)			
Organizational State	*Locus of Dominance*	*Complexity of Assemblage*	*Phases of Unit Interaction*	*Empirical Example*
A	High	Low	Low	Automatic oven timer
B	High	High	Low	Calculator doing addition
C	High	High	High	A servomechanism
D	High	Low	High	A light switch
E	Low	Low	Low	Using a digging stick
F	Low	Low	High	Weaving on a hand loom
G	Low	High	High	Assembly of a prefabricated dwelling
H	Low	High	Low	Hand-operated switch in a railroad tower

dure for deriving the organizational status revealed by the variables of the Man-Artifact Component of Technology, and Table III illustrates procedures for identifying the total organizational status of the three components of a zone.

In assessing an empirical situation, the procedure demands that the three variables that make up a component be measured. For treating our simplified examples, measurements, for reasons given above, are only in polar terms. In collecting data, at least two independent observers should make the assessments. In field tests of the model, we used assessments made by trained operators of the systems being studied, as well as separate measurements by the two researchers.

The specific criteria for the variables' characteristics together with polar scalings of their empirical meanings were given in the descriptive listing of variables above. In deriving the analytical pattern for the variables of the Man-Artifact Component of the Technological Zone in Table II, for example, organizational state A, which coded high, low, low for its constituent variables, is empirically represented by an automatic oven timer. The value assignments were made on the following basis:

Table III. Organizational States Possible for the Three Components of the Social Organizational Zone

	Analytical Pattern (value loadings assigned)			
Organizational State	*Communications Component*	*Domain Component*	*Legitimacy Component*	*Empirical Example*
A	High	Low	Low	A mass aggregate; as in a sports event
B	High	High	Low	Praetorian Guard in the Late Roman Empire
C	High	High	High	An industrial state
D	High	Low	High	Germany in World War I
E	Low	Low	Low	Primitive hunting band
F	Low	Low	High	The Tokogawa Shogunate
G	Low	High	High	Imperial China ca. 1820
H	Low	High	Low	Villages in the Late Ottoman Empire

Locus of Dominance variable was assessed as high since it is an automatic system designed for a single purpose, and this is our requirement for assigning high value to this variable. In similar fashion, low values were assigned to the Complexity of Assemblage variable, there being only one unit involved in this case. Finally, low values were assigned to the Phases of Unit Interaction variable since the oven timer's action is immediate and totally completes the task—*i.e.,* turns the oven off or on.

The other seven organizational states in Table II were illustrated by examples derived in the same manner. Each of these eight combinations represents the organizational state of a single component for the three variables that constitute it.

The three components that make up a zone, once each component is derived from the variables, also exhibit eight states of organization. We list and illustrate these in Table III. Table III illustrates the forms of component organizational states using managerial examples as illustrations. A component is assessed as exhibiting a high or low value from averaging the values of its constitutive variables. Thus, if the three variables measure high, high, low, the component is assessed as high. The empirical examples show what the particular forms of combination mean in behavioral terms. Thus, for example, organizational state F—low, low, high—has the Tokogawa Shogunate of Japan as an illustrative case. In this instance, the three variables that make up the Communications Component were all coded low. These values were assigned because in the case of the Locus of Authority variable, power was shared in a feudal manner between the Shogun and the local feudal lords. The Channels of Communications variable also codes low because of the Shogun's policy of keeping the provinces isolated and the nation as a whole separate from the rest of the world. The third Communications Component variable—Complexity of Linkage—is also coded low, reflecting the lack of effective centralized instrumentalities of control. Since all of the variables were coded low, the entire component is coded as low.

In the Domain Component, the variables of Size-Quantity and Concentration, code low, reflecting the localized nature of the feudal estates and the lack of integration beyond kin ties within the local estates, but which include various non-kin, economically vital retainers within its boundaries. The Heterogeneity variable reflects a growing occupational differentiation in a burgeoning commercial society and tends to code high. The average of the values for the component as a whole, however, is coded low.

Finally, in the case of the Legitimacy Component, the Boundary of Authority and Social Control variables code high, reflecting the attempts of the Shogunate to claim total jurisdictional power over all aspects of life and enforcing these with a secret police and a palace hostage system for the major feudal lords who might be contenders for power. The Locus of Validity variable, however, is coded low, since the Shogun could only claim legitimacy as his due for the historic role of his clan in bringing the disastrous period of civil wars to an end. Again, on averaging the variable values, the component is found to express a high loading.

Applications of the Model

In applying this model to system-state assessment, the first step is to code the current (or historical) situation along the lines indicated in the previous section. This determines an *initial organization state.* In all cases the aim is to assess the state of a zone and in most cases both the technological and the two managerial zones (considering the political as part of management). The analytic pattern that the components take in this synoptic view of the system then permits us to suggest the other seven possible arrangements available to the initial organizational state. In cases where the initial state is to be changed and the characteristics of the "desired state" can be specified in terms amenable to our model, then that state, necessarily one of the alternatives available, may be achieved through the manipulation of the appropriate variables indicated in Tables II and III.

The seven alternative organizational states thus form the basis for the composition of *scenarios* which specify which variables require changes of value in order to accomplish the changes in organizational conditions necessary to create the *desired state.*

The term desired state is not meant in a normative sense, but refers to any other organizational condition which differs from the present. It is desirable in the planning sense as being a chosen *target* of change. For analytic exercises, the model, as developed for presentation in this work, provides seven alternative organizational states that differ from any given current situation and these form the basis for the composition of alternative *scenarios* which specify which variables require changes of value in order to accomplish a transition to the target organizational condition. These scenarios permit the evaluation of policy decisions through comparison of before and after states of the system following deliberately introduced changes in loadings of variables.

The model, thus, permits an assessment of the relative *gap* between the current situation and some desired state. The wider the gap, the higher the costs to society in moving to some future state, and the more complex and difficult the strategies of such a transition. This permits the planner to evaluate, in broad terms, what goals of society are most realistically attainable, given its resources (technological, managerial, and political). (The Japanese and Chinese cases illustrating success and failure to achieve a goal of massive technological change are given in chapters 5 and 6.) Table IV indicates the number of variables requiring change in order to make a transition from one organizational state to any of the other seven potential states.

The scenarios or projected outcomes of manipulation of variables also permit the evaluation of policy decisions through comparison of before and after states of the system following deliberately introduced variable changes. Finally, the scenarios permit the cross-cultural comparison of system-states since our model concerns organizational patterns rather than interpretations of identity of particular cultural items.

Here, again, it is necessary to stress that the procedures in creating such organizational scenarios which illustrate processes of change are dependent on a very simplified set of circumstances.

Only eight organizational patterns are possible when only high and low measures are used for variables. The schematic nature of such scenarios no doubt would rule out effective use and testing of the model for many research situations. However, the principles of developing scenarios from variables that are scaled in a more complex manner would remain the same.

In addition, for the purposes of exposition, the embarrassing issue

Table IV. Number of Variables Requiring Manipulation to Change a State of Organization

Initial State	*Number of Variables Requiring Change to Achieve the Following States*							
	A	B	C	D	E	F	G	H
A	–	1	2	1	1	2	3	2
B	2	–	1	2	2	3	1	1
C	2	1	–	1	3	1	2	1
D	1	2	1	–	2	1	2	3
E	1	2	3	2	–	1	2	1
F	2	3	2	1	1	–	1	2
G	3	2	1	2	2	1	–	1
H	2	1	2	3	1	2	1	–

then surfaces of how to discriminate between a low, medium, or high loading (not to mention points falling in between these), and we are not prepared at this time to explain by what operational procedures different observers or coders can make more truly sensitive discriminations. No doubt we have only partially, at best, been able to avoid the pitfalls of this critical issue in the sophisticated use of this model by the stratagem of choosing what we believed were situations where the data concerned technologies and social institutions for which the loading of variables did not fall too often into painfully intermediate categories.

It should be acknowledged here, once again, that our use of variables, components, and zones involves the acceptance of a principle of *averaging* the constraint values of variables in order to assign a summary loading for components and zones in the model. How legitimate is this procedure? The key assumption is that all the variables only refer to degrees of constraint inherent in a given organizational format and *not* to the entities or conditions themselves. Therefore, all degrees of constraint expressed by any variable equally act as "forces" operating on the organization format. Hence, variables cannot be weighted in any hierarchy of importance. Any shift of value for any variable will have equal effect on the total configuration. To treat real events of social change in the model (next three chapters), it is necessary to reach our "zonal" level of analysis. The variables cannot be utilized by themselves to elicit a system-state for a zone in an additive manner. They must be used first to describe components, which are the minimal units of analysis, and components must be treated as basic units describing zones. The interaction of all three zones is used to model macro-sociocultural conditions such as our Japanese and Chinese case studies represent.

part II

Three Case Studies

This section contains three pilot studies wherein we applied the model to empirical and historical cases that range from specific complex technologies to whole societies. The first study concerns the analysis of the impact of technological change in a modern, highly sophisticated technological system—an air traffic control system. This exercise in analysis, limited to the Technological Zone of the model, serves as an introduction to the operational procedures when applying the model. In the subsequent two chapters the full range of zonal interaction—technological, managerial, and political—is displayed in the analysis of two well-documented cases of large-scale, non-Western societies undergoing change, namely, Japan and China in the late nineteenth century. The analysis of the Japanese case history, while exemplifying organizational change in all three zones of the model, deliberately stressed those features of techno-social change affecting the Managerial aspects. The Chinese case, while equally comprehensive in zonal treatment, was chosen to emphasize the policy-making features of the Political Zone. Taken together, the Japanese and Chinese cases illustrate differential responses to very similar social and technological impacts stemming from contact with European cultures.

The three pilot studies are illustrations of the range of subject matters that can be illuminated by the application of the model. Firstly, as the technological case study demonstrates, it has utility as a diagnostic instrument. In assessing specific technological or social institutions, the model is capable of describing and measuring the degree of viability of the existing state of organization of the activity. It can also predict the effects on technological or social institutions of proposed innovations. At a minimum, it can diagnose which specific components must undergo change and to what magnitude to effect desired changes.

Secondly, as suggested by the Japanese and Chinese studies, in treating total social systems (all three zones), the model has certain implications for applied social science, especially in evaluating and predicting the outcome of a policy decision involving organizational change.

But more important than these practical applications of the model are the scientific implications for generating hypotheses and theories about the processes of change. This feature of the model forms the subject matter of chapter 8, wherein the general rules of organizational change and their theoretical implications are set forth.

4 Case Study 1: Assessment of Technological Responses to Change: Air Traffic Control

The principal emphasis of this chapter is to demonstrate how the model can be used in the analysis of a single zone: the Technological Zone. Even though the total model is designed to treat macro-social and cultural phenomena, a segment of the total model, a single zone, can be treated as a universe of discourse. In such a case, as will be illustrated in this analysis of a complex man-machine system in a modern airport, the model has special utility as a diagnostic analytic tool for the assessment of performance characteristic of the current state of organization of a technological activity as well as for predicting performance that will result from proposed changes. The properties of such zonal analysis are not limited to the technological zone of the model but are equally present in treating the other two zones.

The selection of a radar-approach landing system at an airport was due to the following considerations: Firstly, it represents a complex interaction of men and machines. Secondly, the system has recently undergone change and continues to be subject to further changes. Thirdly, it constitutes a discrete, definable, bounded entity making the investigation of its total parameters of action possible. Fourthly, field work could be carried out rapidly and effectively because the data were easily acquired by direct observation, and knowledgeable and cooperative informants were at hand.

A review of the literature dealing with modern complex technological systems would indicate that technological assessments aimed at analyzing "effectiveness" or at suggesting changes to produce desired results have been subject matters in the normal province of engineers and technologists or of systems analysts like Ackoff (1964), DeGreene (1970), Rau (1970), and Shinners (1967). Analytic models and the data used as inputs have therefore tended to stress the performance characteristics of men and machines in a technological, motivation-psychological frame of reference on the one hand, or an economic or administrative one on the other. When social scientists have involved themselves in such studies, they have tended to stress such issues as social relations, managerial modes of organization, or attitudes and values generated by work conditions. In other words, a common frame of reference whereby these various distinctive study orientations can be unified in a nonadditive way does not exist. Attempts by interdisciplinary approaches to produce a unified theoretical orientation are, when examined, really additive in type, since the concepts and operational procedures used by each discipline is a product of its own subject matter and there is no higher order of abstraction by which each discipline's inputs can be translated into a common theoretical language.

Research Methodology

An empirical demonstration of a process model requires the choice of a situation where change has occurred and is still occurring. Certainly, models exist in abundance which consider the effects of change over time, but they work backward from a given point and compare changes against some achieved state of organization. Other models (note system science, or technological forecasting literature) exist that project organizational changes in an extrapolative mode. However, neither the retrospective or projective models developed to date do both kinds of analyses in the same frame of reference, using the same classes of data and same analytic procedures.

Therefore, for our demonstration we selected a situation where both retrospective and projective analyses are carried out. In the air traffic control example we have collected data required to analyze the operations of this man-machine system at two points in time. These time frames represent the tracon room's organizational configuration before and after important technological innovations were introduced. We have, in addition, certain information about plans for instituting changes in the system in the near future. The analytic procedure, then, yields an interpretation of the effects, organizationally, of past technological innovations on the present man-machine system and projects the organization consequences of possible further technological changes.

These materials fall in the model into three time frames: (1) the past (1968), (2) the "present" [representing the date of field research] (1974); and (3) the future (1980?). The specific time periods to be utilized in applying the model, therefore, are derived from a consideration of the specific character of that which is being studied. The model, however, can also be used for treatment of a single time period in a more limited descriptive analytic manner, though it is primarily designed to analyze situations of change over time.

In the course of study of the air traffic control system, we discovered that the operational characteristics of the tracon room were strongly effected in organizational responses by the level of activity in aircraft arrivals and departures. Treatment of such a condition illustrates the model's ability to treat dynamic variability in structural responses. In other words, we do not have to deal with any ideal or stable type of identity in order to apply the model.

In the tracon room, our observations indicated there were two states of activity that conditioned organizational responses: (1) "normal", that is, defined operationally as a situation when a minimal standard crew, as defined by FAA regulations, can handle the air traffic; (2) "peak", defined operationally by conditions when there is a need to augment the minimal standard crew by additional units to handle increased air traffic.

Field Work and Data Processing Procedures

Before describing the actual air traffic control radar approach system, a review of field procedures and the manner in which data were translated into the variables utilized by the model is called for. The first stage of data gathering in the study of a complex technological activity should be classically ethnographic in character. Observations of the technological activity under investigation were made under all possible circumstances and situations. Key informants, who were actively employed in performing assigned tasks, gave their own versions of operational procedures. In this case, the two investigators viewed the activities independently of each other and then compared observations to determine if similar or dissimilar interpretations were being achieved concerning the observed behaviors or conditions. Similarly, two key informants were asked independently to describe and explain these operational procedures. If conflicting or confusing views emerged, the cases were discussed until all disputed issues were reconciled or reinterpreted.

Since a technological activity, such as air traffic control, may display a changeable pattern of response to different situations, *i.e.*, numbers and types of aircraft arriving at various intervals and under different weather conditions, only a determination of the full range of responses of the system would permit an accurate and relevant analysis. This is equivalent to determining the ranges of organizational responses made by a society as it carries out, or responds to, the variety of events occurring over a functionally significant time period. In this manner, we matched organizational behaviors to the specific contexts of distinctive activities and situations and thereby avoided creating an ideal type or static model of organization.

This body of concrete objective and subjective data amassed in a recognizably traditional social scientific research mode and expressed in traditional descriptive terminology now had to be translated into the variables of the model. Utilizing the definitions of the variables (found in chapter 3), the two investigators analyzed the activities and gave them coded measures. Then the two key informants who had worked with us were asked to make their own independent judgments as to the value loadings of these same activities. This required that the informants had to be oriented as to the meaning of the variables and how to use them. It proved to be a surprisingly easy task. Apparently, technologically oriented and well-experienced operators readily perceive how their behaviors can be measured by the type of variables we developed.

Afterward, the two sets of codings, those of researchers and informants, were compared and if they were significantly different, a discussion of the reasons for this differential in perceived values was initiated and resolved when both groups agreed on an assigned value for the variable. This represents a Delphi-like approach which was applied in a somewhat informal manner. Obviously, in more difficult circumstances, where strong disagreements persist, only a full-fledged Delphi procedure would suffice to resolve discrepancies. However, we believe that despite the crudeness of this initial effort at coding, it can nevertheless be defended on the ground that the judgments of well-trained scientific observers and those of highly experienced technicians, where they closely agree, can be trusted.

The Tracon Room, Air Traffic Control Center[1]

The tracon room studied is a facility of the Federal Aviation Agency, which is located in the control tower of the Portland International Airport. It is one of a number of such facilities associated with all major airports throughout the country, each of which is responsible for the control of airspace within prescribed limits directly surrounding a given airport. At the Portland Airport the airspace is roughly a circle thirty miles in radius centering on the airport itself.

The facility is located in a room in the tower directly under the control deck from which direct visual control is exercised during the actual landing and takeoff operations. It is the function of the tracon room to guide aircraft through the outer portions of the airspace and then hand over the last phase of landing to the upper-deck control center of the control tower.

One wall of the room consists of electronic control consoles, four in number, each of whose central feature is a large radar display screen. An air traffic controller and his assistant are seated in front of a radar unit and from here control an assigned section of airspace. In addition, there are radio and computer printout terminals prominently displayed on the walls. The computer terminals are the primary source of data concerning scheduled aircraft arrivals and departures (but, as we will see, other *un*scheduled flights must be handled, as well).

The typical routine of the aircraft control procedure is as follows: A pilot first files a flight plan with the FAA. This material is sent to a central computer located in the Seattle Regional FAA Center where the basic data (time of departure, type of aircraft, estimated time of arrival, etc.) are recorded on tape and printouts of these data sent to

[1]An excellent survey of literature on CAA and FAA air traffic control research is found in Parsons (1972: 283–302) and indicates that the emphasis has indeed focused on improving the men or machine components of the system, which could only provide partial solutions to the overall problem.

the appropriate airport control stations along the proposed route. The computer printout terminal in the tracon room is the source of this flight data. A slip of printout paper with this data is then handed to the controller within whose sector of assigned airspace the plane will be flying. At the Portland Airport tracon room in normal, nonpeak traffic periods, the airspace (360°) is divided into only two sectors, at peak traffic periods, there can be as many as four air traffic controllers operating at one time as the crowded airspace requires more division of labor. The controller's duties consist of assigning each aircraft airspace by regulating the altitude, speed, and direction of the aircraft. Thus, the controller must exercise a very complex and continuously shifting awareness of all the coded data on his radar scope that refer to each aircraft and also to their relation to natural topographical danger points. At rush periods there can be several dozen aircraft in each section, each moving at different speeds, altitudes, and directions.

The above descriptive summary represents the current state of organization of the tracon room facility. From this point in time we move backward and forward in generating scenarios, which contain the analyses of organizational states past, present, and future.

The past, represented by the situation in 1968 (data for which time period was gathered in an earlier ethnographic exercise), exhibited an organizational configuration unlike the present in certain easily noted ways. The primary feature of the system at that time was an emphasis (in comparison to the present) on the human component of the man-machine complex. No centralized computer facility was available to give flight plan information and the older radar sets were not equipped to receive and display automatically signals from aircraft giving their speed and altitude. This meant that the controller had to contact aircraft by radio and carry out all sorts of judgmental behaviors in monitoring and controlling individual aircraft. In addition, all flight plan information was received by telecommunication from departure points and circulated to controllers in handwritten form.

Now that we have traced the salient features of the tracon room's composition and activities, we will code the ethnographic descriptive data for the *past* system state, variable by variable, in order to generate the class of data needed for the construction of scenarios which illustrate how the model works. In the model, this time period appears as an organizational state described and measured by the coding of the three components of the Technological Zone.

Past Organizational State of the Tracon Room

The Man-Artifact Component, when the three variables are coded, averages as *high.* This evaluation was made in the following manner. The Locus of Dominance variable is coded as *low* because of the greater involvement of human behaviors in the task as contrasted with machine components (as compared to the present). The Complexity of Assemblage variable is coded as *high;* there is a large assemblage of interacting units (whether these are men or machines is not relevant for coding this variable). The Phases of Unit Interaction variable is also coded *high* because of the linked series of actions that the controller must follow procedurally in guiding incoming aircraft.

The Task Component averages *high* because its variables have the following codings. The Serial Characteristic variable is coded *high,* as task completion requires an ordered set of action over time. The Operations-Output Relations variable also codes *high* since these operations of the tracon room are dispersed over time and space; that is, what is being controlled, namely aircraft, are coming from all points of the compass. The Output Form variable codes *low* as there are often random inputs since not all flights are scheduled.

The Setting Component averages *low* as a result of following codings. The Setting Structure variable codes *low* since there are only limited controls exerted over private unscheduled aircraft arrivals. The Locus of Input variable, however, codes *high;* obviously, aircraft arrive from all points of the compass. The System Autonomy variable codes *low* as well because the tracon room operations are very self-sufficient and do not depend on outside units to assist them in the specific control of aircraft once these enter into its airspace.

In sum, for the past organizational state of the tracon room, the coding of the three components of the Technological Zone of the model represents Organizational State B (Table III). Before applying the model for an analysis of change, it is necessary to contrast this organizational state with a more current situation (1974).

Current Organizational State of the Tracon Room

We will see from the coding of the variables of this time period that the current state of organization also represents Organizational State B, that is, a component pattern of *high, high, low.*

The Man-Artifact Component's variables all code *high.* The Locus of Dominance variable is coded *high* since the whole system is dominated by a single purpose, the effective control of aircraft. The Complexity of Assemblage variable is also assigned a *high* value since the system consists of an assemblage of units: radars, computers, radio links to planes and other control centers all interacting in the context of the task. The Phases of Unit Interaction variable also exhibits a *high* value assessment since the assemblage of equipment operates in a multiphased fashion, each step being dependent upon the completion of the previous step in the process. For example, in dealing with aircraft departures, the flight plan is sent from the central computer to the controller who integrates the data with that displayed on his radar screen. He then controls the flight until the plane is "handed-off" to the next air traffic control center. Since all of the variables in the Man-Artifact Component code high, the component is assigned a *high* value.

The variables of the Task Component show a *high* loading for the Serial Characteristic; that is, task completion requires an ordered set of actions over time. This reflects the Phase aspects of the Man-Artifact operations and the fact that many aircraft flights are scheduled by the Federal Aviation Authority. The Operations-Output Relations variable also codes *high* since the operations of the control center occur dispersed over time and space—that is, throughout the defined airspace surrounding the center. The Output Form variable, however, exhibits a *low* loading. While there are routines to aircraft traffic control, the system cannot be rigidly specified since random entrances into the airspace (discussed below) and emergencies can and do occur. Thus there must be some room for improvisation within the Task specifications. The Task Component as a whole is assessed as being *high* in its value loadings since the three variables average to high.

The Setting Component average shows a *low* loading. The Setting Structure variable is assigned a *low* value since there are some uncontrolled aircraft entrances into the airspace. Private planes and military flights intrude upon the scheduled flight operations and this opens the system to random inputs. The Locus of Input variable codes *high* since the sources of the inputs—aircraft flights—come into the system from diversified points. The Autonomy variable also codes *low,* the Air Traffic Control Center being essentially self-sufficient in controlling its assigned airspace.

Now that we have established codings in the model for the two organizational states representing the past (1968) and the present (1974) we can conduct an analysis of the process of change and its implication for the future.

A comparison of the codings leads to a conclusion that even though a single variable changed its loading from *low* to *high* in the Locus of Dominance variable of the Man-Artifact Component, the organizational state, as measured by component *averages,* remains the same: *high, high, low.* This implies that something far less than a major organizational shift has occurred in the system as a whole. The attempted change, that of introducing more efficient machine elements in the man-machine balance of interaction, did not, in the end, solve the problem of total system overload—that is, more aircraft coming into the airspace than can be effectively controlled by human operators once the machine components are "swamped". The human operators in 1968 were under severe stress when coping with overload or peak conditions and remain so today, despite obvious improvements in the machine components of the technology. Here is where the model displays its strength in analysis because it is not limited to treating the man-machine components as distinct from *task* or *setting* conditions. In this model, these receive equal attention. In the tracon room airport case, our analysis of the total system (made up of air traffic controllers, navigation aids, and aircraft) indicates that the empirical problems of overworked and overstressed air traffic controllers and overcrowded airways relate most crucially *not* to poor man-machine interfaces, but to policy matters affecting use of airspace, which, in the model, is a "setting component" problem. The low loadings for all three variables of the Setting Component constitute the diagnostic indicators for the system's difficulties. Specifically, the improved performance of the computer-structured flight control characteristics only generated more flights, leading in turn to greater stress on the system since the uncontrolled arrivals of aircraft only increased in volume. In other words, failure to control inputs (modelled by the variables of the Setting Component) wipes out the advantages obtained by improvements in the purely technical components of the system. Therefore, it would appear that any further changes made in Man-Artifact features of the system without corresponding changes in other aspects of the total system is doomed to various degrees of failure.

So far in the analysis of change from the recent past to the present, we have concluded that fundamental organizational changes in the total system did not occur. Now we illustrate how analysis can proceed from this base to predict future conditions and consequences of changes.

Future Organizational States

The analytic and diagnostic procedures so far carried out indicate that changes in the Setting Component are most likely to produce a more effective system. The improved organizational state suggested by the model is a shift from Organizational State B to *Organizational State C (high, high, high),* in other words, a rise in the value of the Setting Component from *low* to *high.* In this instance, the component that measured low in the initial state, the Setting, will require a change in the loadings of only one of its variables so that it will, upon averaging, yield a high valuation. This will then change the component's pattern and reflect a change in the overall system-state.

Among the Setting variables of the current state, one of them, the Locus of Input variable, already measures as *high.* Thus to change the value of the component's average, only one of the two other variables of this component requires a shift in its loading. The manipulation of the Autonomy variable to a higher value would mean that there would have to be greater integration of the local air traffic control center into the national control network. But the fact that local conditions, such as weather, cause a need for autonomy, the manipulation of the Setting Structure variable would appear to be more promising. The value for this variable could change by tightening the requirements for entering and operating within the center's assigned airspace. The disciplined scheduling of private aircraft flights together with their required use of transponder devices (instruments that automatically record altitude and indicated ground speed on the controller's radar screen) would accomplish this end. If these requirements were met, the randomness of aircraft entrance into the airspace would become controlled, resulting in the loading change for the variable that the scenario calls for.

Now we can explain a further feature of the model. It addresses itself to the critical problem of how to choose the best potential strategy of change from among an array of possible organizational formats. The current state of the airport's tracon room is System State B, and we have shown that movement to State C would produce a more effective overall pattern. But what would changes to the other Organizational States of A, D, E, or F imply? Here the consequences of such changes can be modelled in the production of alternative scenarios.

Since some of these alternative states of organization represent *lowering* the constraint measures, it would be a useful exercise to show how the model can be used to project changes in the direction of lesser rather than greater complexity in components or a zone as a whole. Therefore, let us choose Organization State F (low, low, high) as an example since it requires not only lowering values of variables but requires that the value of all three components contrast with the current Organizational State B (high, high, low). It would represent a more radically different set of changes than change to Organizational State C, already treated, where only modification of one variable occurs.

As indicated above, the Man-Artifact Component in the current organizational state shows that its three constituent variables all code high. To change this component's value to low, two of these variables require shifts to low loadings. Since the Locus of Dominance variable concerns the primary purpose of the system, and air traffic control centers are difficult to utilize for other functions, the most probable course of action in changing this component lies in shifting the loadings for the other two variables. The Complexity of Assemblage variable can achieve a lower value by the abandonment of much of the sophisticated electronic equipment now used in their operations, and relying on visual procedures such as characterized aviation practice in the early period of its existence, circa 1920. This change would also lower the loadings for the Phases of Unit Interaction variable. The drastic lowering of the number of flights that would take place would result in each flight having simultaneous phase features.

The Task Component in the new organizational state requires a value shift from high to low. In the initial state the Output Form variable was already coded as low, so that desired changes must be made in either of the other two Task Component variables. The loadings for the Operations-Output Relations variable can only be lowered if the actions it concerns are localized. This, in the case of air traffic control, is impossible to accomplish, so the only viable alternative is to manipulate the Serial Characteristic variable. This could be accomplished through making the temporal aspects of aircraft control more simultaneous—perhaps by making all flights into the airspace automatic through computer control of all aspects of aircraft operations. Such an approach would lower the values of this variable, and permit the component's average to attain its required loading. Unfortunately, for the implementation of such a scenario, the movements of private aircraft and specialized military requirements cannot, in fact, be so disciplined and restricted for a whole host of reasons.

The Setting Component's loading can be raised by the following changes indicated in the scenario for change to System State C, where such a shift is also required.

If one compares alternatives of change from the current Organizational State B to C as opposed to F, it becomes clear that, in practical, substantive terms, C is far superior to F. It requires fewer changes in loading of variables. Further, changes that lower measures in the Man-Artifact and Task Components involve drastic shifts to a more primitive technology in which *fewer* aircraft could be accommodated, and this could not be tolerated as the basic system requirement is that it must handle *more* aircraft efficiently.

The tracon room analysis highlights a very significant point about technological and social change. Changes required to make the Air Traffic Control System more viable, if they are indeed linked to the way in which aircraft are allowed into the system, implicate issues that are by no means of technological nature. They have to do with matters of law, tradition, and ideology. The rights of pilots, who are citizens, to make use of airspace and airport facilities and services are important in current American society, even if they conflict with the interests of major airlines or military needs. Hence, solutions in these areas are political rather than technological in nature. To try to accomplish change in complex systems such as an airport facility thus requires awareness of the need to make a total organizational analysis.

In the next chapters we demonstrate how the model deals with political and managerial accommodations that must be made in the face of drastic technological changes. There the materials used as data are published historical accounts rather than a modern, firsthand field research report. This we hope will serve to indicate that the model can be utilized when either class of information (or even combined types) constitute the data inputs.

5 Case Study 2: Successful Social Organizational Responses to Technological Change: the Japanese Case

This chapter concerns itself with the expositional illustration of the application of the social organizational variables of the model to a case of policy analysis; specifically, the empirical measurement of the success or failure of a classic attempt to introduce industrialism and modernization into a pre-industrial society. The case study is that of the transformation of Tokogawa Japan into a modern industrial state. It will illustrate the powers of the model for social organizational analysis and introduce its potential for monitoring social change.

The model can assess failure, together with the reasons for failure, as readily as it does a success in strategies of change. A failure is defined by the model as a choice of transitional organizational steps that will not result in the degree of organizational constraint required to meet the requirements of technological tasks because the measurements of variables show its constraint features to be less than is necessary to achieve the new organizational format. A success is defined, similarly, as a transition that follows the rules of transformation required for that organizational state to attain another format by changing the values of variables in the manner and to the degree prescribed in the model.

The present illustration of the model's properties shows that eight possible organizational states are generated ranging in their various variable configurations or constraint formats from high to low.[1] Movements of change measured by the variables can be accounted for in any direction: high to low, or vice versa. The organizational states generated by the model can be homogeneously high in all of their components, or homogeneously low, or show a mixed format. Any empirical sociocultural entity (*e.g.,* from a band to a modern state) will approximate one of these eight classes of organization. All change is demonstrated in the model as a change from one organizational state to another of the seven alternative types.

The rules of transformation of an organizational state into any of the seven alternative states are plainly stated in terms of the number of variables that have to be changed, and the direction of change in value loading (high to low or vice versa) is indicated for any strategy of change. Therefore, the selection of an appropriate alternative state, and the following of the rules of transformation to change that state into the desired one, constitutes a *strategy of change.* The causes that initiate a desire or need to change existing technologies or social institutions are not issues requiring analysis in the model. The analysis begins at the point when action has been initiated or specifically proposed. The model then predicts the outcome of these strategic choices.

Clearly, such matters are of interest to the policy planner. The model is intended to permit the evaluation of policy decisions and the derivation of certain principles of policy selection and implementation. Specifically, our analysis of the Japanese case will illustrate the necessity of providing for adequate management of the technology at the proper level of action by an entity having the necessary organizational characteristics (as specified by the model).

[1]Later refinements in coding variables will permit much more precise descriptions of organizational states. Many more than eight organizational formats are generated when the crude polar high and low values are replaced by numerical progressions.

Selection of Technologies and Social Groups for Modelling

Our selection of Japan as a case study was determined in part by the consideration that published data amenable to modelling in our approach were readily available.[2] More importantly, Japan was never a colonial possession of the Western powers, and so the effects of the policies that were adopted were not simply incorporations of cultural traits imposed by an alien society.

Our purpose in this analysis is not to describe Japanese culture, nor is it to consider all phases of the Meiji Restoration. Our only purpose is to show the effects of strategic decisions on policy in the sphere of social responses to particular technological changes.

While industrialism implies a whole spectrum of radical technological changes, we have limited our analysis to three fundamental technological forms: transportation and communications, the steel industry, and military technology. Our selection was dictated by the fact that these three areas were the object of deliberate official governmental policy decisions meant to produce radical changes.

For coding purposes we may operationally limit ourselves to certain selected aspects of broad technological systems. Transportation and communications systems are understood to be systems by which men, materials, and information are moved over space. Military technology consists of classes of weapons systems-individual (rifles), crew served (field artillery or warships), and technical and logistical systems (engineering or medical services). Heavy industry is represented by the steel production complex, in part because many other industrial developments are predicated on its existence, and partially because it represents (for the periods being considered) a very complex form of organization.

The three areas of technology, once selected, conditioned our choice of social groups to be considered as managing entities. Of the numerous social categories that composed Japan in both the Tokogawa and the Post-Meiji periods, our concern was only with those involved in the control or implementation of the three technologies we have specified.

There are three basic social groups of concern in the Tokogawa period: the Shogunate; the Daimyo, or feudal lords; and various mer-

[2]The data utilized in this modelling exercise were collected from the following sources: Gubbins (1922), Honjo (1935), Kobayashi (1922), Latourette (1938), Nitobe (1930), Norman (1940), and Samson (1931).

chant-bankers, in particular, those who amassed great wealth. In the Post-Meiji period, the ruling entity, a Shogunate, was replaced by a centralized state government, ruling in the name of the emperor, and the feudal lords and merchant groups were merged into a monopolistic oligarchy—the Zaibatsu which, while no longer feudal in a territorial sense, did preserve a certain feudal organizational format in their new areas of managerial control in industry and finance.

Assessment of the Technological Zone

In applying the procedures used in the model, we begin by ascertaining the organizational structure of the Technological Zone. The coding procedure requires that the empirical inventory of technology be assembled, in this case from historical sources. Using the variables of the Technological Zone (Table I), these inventories were coded independently by the two authors. The coded assessments of the totality of individual items (a cart) comprising a particular technology (transportation) were then averaged for all of the items of that category. The component loadings for each of the three technologies were then derived from averaging the three variables that make up the component.

The next step was to provide such coded component values for the initial state (the pre-industrial state of Tokogawa Japan). We then proceeded to present the codes for the *desired state* (*i.e.,* codes that represent the organizational format for Western European technology of the time). Finally, we provided the codes for the *achieved state,* that is, the codes for the technological inventory (for the three selected areas) as it existed in 1910. These data are summarized in Table V.

The following descriptive materials form the basis for the coded assessments of the two periods as presented in Table V.

The values assigned to code the technological and social organizational state of Japan are relative to the standard of complexity current in late nineteenth century Europe. Obviously, Tokogawa Japan represents a more complex state of organization than a tribal society; however, our codes are not based on an absolute scale at this point. The high versus low loadings are only provisional, illustrative assessments. The contrast between the initial and the achieved states forms the basis for the analysis of the changes in the Technological Zone.

Table V. Coded Assessments for the Technological Zone

	Transportation		*Military Technology*		*Steel Industry*	
Variable and Component	*Tokogawa*	*Post-Meiji*	*Tokogawa*	*Post-Meiji*	*Tokogawa*	*Post-Meiji*
MAN-ARTIFACT COMPONENT						
Locus of Dominance	Low	High	Low	High	Low	High
Complexity of Assemblage	Low	High	Low	High	Low	High
Phases of Unit Interaction	Low	High	Low	High	Low	High
TASK COMPONENT						
Serial Characteristics	Low	High	Low	High	Low	High
Operations-Output Relations	Low	High	Low	High	Low	High
Output Form	Low	High	Low	High	Low	High
SETTING COMPONENT						
Setting Structure	Low	High	Low	High	Low	High
Locus of Input	Low	High	Low	High	Low	High
Autonomy	Low	High	Low	High	Low	High

Tokogawa Codes = Initial state
Post-Meiji Codes = Achieved state
Desired state (Western European technology) codes high in all categories

Technological Assessments of Tokogawa Japan

Transportation and Communications. The transportation system of Tokogawa Japan was very limited, essentially utilizing man- or animal-powered land vehicles and sails for small coastal vessels. It was the avowed policy of the Shogunate to keep Japan isolated, both externally and internally, with the result that ships were restricted in size to coastal vessels and the interior road system was deliberately allowed to fall into disrepair. Land transportation was confined to horses, carts, and human porters. Accordingly, we have coded these materials as *low.*

Military Technology. Compared to late nineteenth century Europe, the Japanese arsenal was very limited. The professional military castes utilized the sword and bow and arrow, with lacquered leather and metal armor. The peasant levies were usually armed with pikes and an occasional matchlock. The only artillery available was in the form of cannon in fixed fortress positions. Some of the coastal vessels were lightly armed, but none had armored turrets or hulls. We coded these materials as *low.*

Steel Industry. The simple iron foundries available at various localized sites hardly constituted a modern steel industry. The production of most goods occurred in localized cottage industries. These industrial operations are all coded *low.*

Technological Systems of the Desired State

The Meiji reformers sought to effect a radical technological revolution in Japan for the purpose of achieving and maintaining national autonomy in the face of Western European and American imperialism. To accomplish this, their policy demanded that Japan adopt the existing technology that characterized Western Europe at the time. In order to compare the actual achievements of the Meiji Reform against their self-set standards, we included a short description and coding for the European technologies in the three relevant areas.

Transportation and Communications. Two primary technological developments were of importance in the transformation of Western transportation and communications systems in the nineteenth century: the introduction of steam power to land and sea transportation, and the development of telegraphic and cable communications. Both of these innovations radically transformed the capacity and time rates involved in the movement of men and materials and information over space. Accordingly, we coded this area as *high.*

Military Technology. The developments of the transportation and the steel industry permitted the rapid transformation of military technology and the subsequent political predominance of European powers. Field artillery, rifled personal weapons, and machine guns multiplied fire power, while railroad or steamships permitted the rapid concentration of troops or ships at selected points. The improvement of communications and the rapid development of specialized technology required the establishment of trained professional staffs and technical services, *e.g.,* engineers, as part of the military establishment. These data are coded *high.*

Steel Industry. The industry is coded as a total system, that is, as including rail transportation of the raw materials to the blast furnaces, the rolling mills operations, and the production of the finished ingots that are then utilized in various ways by other associated industries. The complexity and diversity of these factors led us to code this aspect of technology as *high.*

Technological Systems of the Achieved State

We have selected 1910 as the year for assessment of the Post-Meiji period of technological development in Japan. This is some forty years after the introduction of the proposed changes and yet precedes the First World War, which, while of little consequence technologically for Japan, did have profound effects on the technological structure of Europe. In addition, the forty-year period provided enough time for the introduced technologies to become part of the established social conditions of Japan, as well as permitting the workers and managerial personnel to achieve a certain proficiency in applying the technologies.

Transportation and Communications. Steam railroads were well on the way toward becoming a national network. Steamships were being built in Japanese naval yards and used in regular worldwide service. Internally, aside from the railroads, the road system, while in better repair, still operated much as in Tokogawa times. A national communications network of telegraph lines and cable systems tied Japan to most parts of the world. We coded these conditions as *high.*

Steel Industry. A well-established industry existed in 1910; it formed the basis for other associated industries utilizing steel in their operations—such as shipbuilding, armaments, railroad building, construction, and so forth. We coded this industrial establishment as *high.*

Military Technology. The Japanese military establishment was completely revised. A national standing army was established and equipped with modern weaponry—rifles, machine guns, and field artillery. The appropriate technical and logistical services were also introduced. In the naval field, a modern fleet of steam warships and navy bases was established. The appropriate command structures were also developed on Prussian (for the army) and British (for the navy) lines. The prowess of the new forces was adequately demonstrated in the Russo-Japanese War of 1905. We coded these developments as *high.*

Assessment of the Managerial Organization: Tokogawa Period

The principal focus of this chapter is on the social organizational responses of Japanese society to the changes in technology that we have described. As indicated in the introductory material above, these responses are of two sorts: managerial and political. Managerial responses are those that occur as a direct consequence of carrying out specific technological activities.

In specifying the social groups that are of concern to us in modelling social change, we noted that three groups were predominant in their managerial roles—the Shogunate, the Daimyo, and certain merchants. In this section we shall present coded evaluations of the managerial roles that each of these three groups played in controlling the technology of the Tokogawa period. This set of codes will thus constitute the initial state of managerial organization. Following this presentation, we shall provide similar codes and descriptive materials for the reorganized controlling social entities of the Post-Meiji period that constitutes the achieved state. The comparison of these two organizational states will provide, analogous to that in technology, a means to analytically ascertain the successes and failures of the organizational reforms in the Managerial Zone.

THE COMMUNICATIONS COMPONENT

Locus of Authority Variable. The Shogunate, with respect to transportation, sought to isolate the individual feudal lords from each other, and the country as a whole from the rest of the world; accordingly, roads were neglected and left in disrepair, and the size of ships that were constructed was limited by law. In terms of military action, the Shogun had only a feudal right to call upon loyal clans and his own retainers. Though the Shogun built a cannon foundry, so did other Daimyo. Thereby the Shogun had no technological advantage over the powerful feudal lords (code = low).

The Daimyo had among their number great feudal lords at least equal to the Shogun in actual managerial strengths. Indeed, since most productive enterprises were localized, they possessed much more practical authority within their territorial holdings than did the Shogun (code = high).

The merchants, on the other hand, essentially depended upon cooperation and collaboration with the Daimyo to enter certain technological sectors. Their primary involvement with a technology of concern to us is in the field of transportation (code = low).

Channels of Communications Variable. The Shogun's primary role was political so that apart from managing his own clan's territories, the technological impact of his policy was actually to limit channels of communication of others (code = low).

The Daimyo in their narrower sphere nevertheless had total responsibility for the management of affairs of what were for the greater lords mini-states. All transportation, industry, and military affairs were directly controlled. The larger Daimyo were concentrating industrial as well as military power which resulted in evolving numerous channels of communications among the complex components they were orchestrating (code = high).

The merchants, as a commercial class, did have certain trans-provincial activities (*e.g.,* grain transportation), but for the technologies concerned were hardly important organizers as compared to the great Daimyo (code = low).

Complexity of Linkage Variable. Just as the political interests of the Shogun dictated that he weaken his opposition by limiting their channels of communication, the same interests forced him to attempt to lower the complexity of linkage potentials of other competing organization groups lest they combine to overthrow his rule. Actually, from an industrial standpoint, the essential local cottage-industry nature of most productive processes ensured that few linkages would occur on this basis. Military needs, however, had high priorities and did press the Shogun to seek to maintain a complex set of links with the military aristocracy. But the feudal nature of linkages between the Daimyo left much to be desired in providing a sure and effective fighting force for the Shogun. Feudal warfare itself exhibits hardly any of the phased operations characteristic of modern warfare (code = low).

Since the Daimyo were true managing entities in their territories, the more powerful ones in fact demonstrated complex formats for linking their many enterprises together. Their concerns might not have been as broad as the Shogun's but they certainly involved iron production, military preparedness, and transportation facilities (code = high).

The merchants, while engaged in nationwide enterprises of a commercial nature, hardly resembled the greater Daimyo in their involvements in other technologies. They were more like the Shogun in that they too had to manipulate the Daimyo to achieve their ends in a more indirect manner (code = low).

THE DOMAIN COMPONENT

Size-Quantity Variable. The Shogunate utilized the same technologies and to the same scale as the Daimyo and merchants. Industrial activities were dispersed among the various Daimyo. In military affairs the Shogun was barely able to muster a larger force of loyal clans than his potential rivals (code = low).

The Daimyo resembled the Shogunate in organizational pattern for this variable. While the simple productive technologies were carried out in the feudal estates rather than "nationally", this advantage of the lords was offset by the Shogun's greater military potential (code = low).

The merchants, for the three technologies, must code low since as a commercial class they were not involved in military organization and even when they had some managerial concerns in industrial production, it was in collaboration with the Daimyo (code = low).

Heterogeneity Variable. The Shogun shared with the other two groups the same kind of technologies. The ships, carts, and human carriers reflected a simple transportation technology. The military organization contained only two arms—infantry and cavalry, neither of which was complex in internal organization. The manufacture of iron was on a craft level and output of cannon or muskets was very limited (code = low).

The merchants, similarly, participate in a technological setting characterized by simplicity and homogeneity rather than heterogeneity (code = low).

Concentration Variable. The Shogunate was only vitally concerned with maintaining order and securing revenues so that military and coercive aspects of technology were of direct concern. It would appear that the Shogun would have to score high on this variable since those he would control and tax were found dispersed in space. But in the mid-nineteenth century the Shogun had limited success in such matters. The Shogun was most effective in managing the technological activities of his own clan. The Shogun's national army could only be assembled for limited periods. His own forces were a palace guard and the retainers of his clan (code = low).

The Daimyo, representing a feudal entity, were localized rather than dispersed in their managerial activities. They indeed had managerial tasks involving all three technologies, but they were narrowly focused on their own clan territories (code = low).

The merchants, because of certain specialized concerns which involved transport of goods between provinces, did have an important set of widely distributed activities they could effectively carry out. In this feature they contrast with the Daimyo, who lacked such interests, and the Shogun, who indeed had many interests involving the whole country but had no effective organization of his own (code = high).

THE LEGITIMACY COMPONENT

Boundary of Authority Variable. The Shogun claimed total control over all the affairs of the empire but in actuality lacked the means to enforce it with consistency. Much of the techniques used were of a negative nature, *i.e.,* an internal passport system to prevent contacts between the Daimyo or the ban on contacts with foreigners. No real attempt was made to foster industry. The Shogunate had deteriorated from the Hideyoshi beginnings when its many areas of jurisdiction were effective to a point where it had lesser areas of jurisdiction and was less effective even in the remaining ones (code = low).

The Daimyo more effectively could exercise complete control over the many affairs in their territories by their ability to collect taxes and initiate and manage some industrial activities. The powerful Daimyo could outshine the Shogunate in several technological aspects in part because their boundaries of authority were successfully asserted in practice due to their tight controls locally. They could use feudal loyalty in this localized format effectively, whereas the Shogun could not in a similar feudal context, nationally (code = high).

The merchants in this feudal order could assert no claims to any formal areas of jurisdiction even when they managed a transportation or productive activity (code = low).

Social Control Variable. The Shogunate, as noted, could manage national affairs only by restrictive or coercive techniques. Despite all such attempts, the Shogun could no longer muster the military or economic force to counter the growing resources of the Daimyo and merchant-bankers. No attempt was made to compete with the nobles in industrial or communication development at this time (code = low).

The passport and palace hostage system tended to keep the Daimyo in a precarious power balance *vis-à-vis* the Shogunate, politically. But in the management of their own enterprises, their techniques of social control were most effective (code = high).

The merchants exercised some aspect of social control indirectly by giving or withholding loans to the Daimyo they wished to influence. They had no coercive powers such as military or police agents to enforce their demands (code = low).

Locus of Validation Variable. The claims of the Shogunate to total power in the realm were ambiguous in their effects since the local Daimyo had their own police and military force, transportation facilities, and industrial activities. Managerially, the Shogun's position did not permit him to use his idealogical claims to leadership in practical psychological affairs. He was most effective in asserting his personal authority over his own clan in such matters (code = low).

The Daimyo who had enormous local executive powers still depended on personal factors of family status and warlike abilities to maintain authority over technological enterprises (code = low).

The merchants, of course, in feudal Japan could make no claims to abstract principles in support of their financial control over certain enterprises (code = low).

These codes are summarized in Table VI.

Assessment of the Achieved Managerial State: The Post-Meiji Period

Having coded the initial state for the managerial organization of the late Tokogawa period, the next step in the analysis will be to elicit the codes for the Post-Meiji managerial activities. This is done in a manner similar to that followed in the presentation of the Tokogawa data. Here we must note that in the Post-Meiji period the number of classes concerned with management has been reduced from three to two and their positions radically changed. The Shogunate was abolished and a centralized state government was instituted to rule in the name of the divine emperor. The feudal Daimyo had their clan lands expropriated and were reimbursed for their loss not in monies but in government bonds. The former Daimyo and the merchant-bankers were merged into a new managerial entity—

Table VI. Coded Assessments for the Managerial Zone

	Tokogawa Period			*Post-Meiji Period*	
Variable and Component	*Shogunate*	*Daimyo*	*Merchants*	*Central Government*	*Zaibatsu*
COMMUNICATIONS COMPONENT					
Locus of Authority	Low	High	Low	Low	High
Channels of Communications	Low	High	Low	High	High
Complexity of Linkage	Low	High	Low	High	High
DOMAIN COMPONENT					
Size-Quantity	Low	Low	Low	Low	High
Heterogeneity	Low	Low	Low	High	High
Concentration	Low	Low	High	High	High
LEGITIMACY COMPONENT					
Boundary of Authority	Low	High	Low	Low	High
Social Control	Low	High	Low	High	High
Locus of Validation	Low	Low	Low	High	High

Tokogawa Period = Initial state
Post-Meiji Period = Achieved state
Desired state (Western European managerial organization) codes high in all categories

an oligarchy that, in cooperation with the central government's bureaucracy, initiated the technological changes we model below. The Post-Meiji period thus has two managerial entities to consider: the government bureaucracy and the oligarchic Zaibatsu.

THE COMMUNICATIONS COMPONENT

Locus of Authority Variable. The central government took upon itself the task of initiating through its own policy decisions the establishment of a national railway, telegraphic and cable systems, and steamship lines. By expropriating the lands of the feudal Daimyo, they blackmailed the former feudal managers into becoming a national managing elite for the new technologies. European models were adopted wholesale but in all cases the Japanese kept jealous and exclusive control over all aspects of the various burgeoning enterprises. Thus the establishment of the steel industry follows this pattern. With the adoption of the Prussian military staff model for the army and the British model for the navy, the central government firmly held the locus of authority in its own hands. Nevertheless, except in the case of managing the military aspects, the direct management of technologies was in the hands of the Zaibatsu. Once the various industrial enterprises were established, the government tended to leave further development of the given technology to private sources (code = low).

Thus, the Zaibatsu emerged as paramount when referring to the direct management of industrial techniques, including the armaments industry (code = high).

Channels of Communications Variable. Though the central government concerned itself more with policy formation (hence a political function) than with direct management, its general adoption of a European mode of bureaucracy was responsible for an enormous increase in channels of communication available to these new technologies. Unlike the Shogunate, its policy was to develop the technical communication means of unifying the country. Hence, it fostered road and bridge construction, rail and steamship transport, and telegraphic communications. The bureaucratic interface between government and Zaibatsu was marked by increased channels of direct communication (code = high).

The Zaibatsu, as a result of their adoption of a bureaucratic mode of operation and control in their management of the new technologies, match the central government in complexity (code = high).

Complexity of Linkage Variable. The central government established complex linkages within and between all three technological areas via the state bureaucracy, which was an integral part of the machinery of organization of new technologies. Since these new technologies were based on the forms of Western European models, the government bureaucracy exemplified all the complex patterning of an industrialized state (code = high).

The Zaibatsu, as already described, followed the state's assumption of a complex bureaucratic organizational format (code = high).

THE DOMAIN COMPONENT

Size-Quantity Variable. Since the central government used its resources to encourage the Zaibatsu to develop the three technologies, its managerial role is somewhat indirect. Only in its management of the armed forces is a high managerial coding possible (code = low).

The Zaibatsu in 1910 controlled and managed a modern rail, steamship, and steel industry which in terms of size and quantity characteristics obviously must code high (code = high).

Heterogeneity Variable. The central government even with secondary involvements in establishing new industries, nevertheless, had to contact foreign suppliers of resources and technological equipment. In such dealings with foreign governments and great private corporations, the government had to develop large numbers of agencies, both administrative and technical (code = high).

The Zaibatsu likewise show an enormous increase in heterogeneity over the Daimyo feudal format as a result of specialization of function and division of labor involved in establishing new technologies along foreign European lines (code = high).

Concentration Variable. The national and prefectural bureaucracies were dispersed, effectively leaving no administrative vacuum as they replaced the locally autonomous feudal system. The government made its presence known throughout the land and this was largely due to the technological assists of the new rail and telegraphic communication links. The army and navy also permitted the imperial government to operate in a dispersed manner but with great efficiency (code = high).

The Zaibatsu controlled various technologically based enterprises whose nature it was to be dispersed in space. The steel industry required the importation of huge quantities of raw materials, and elaborate rail and shipping facilities were now required to bring the various operations into some integrated pattern (code = high).

THE LEGITIMACY COMPONENT

Boundary of Authority Variable. Since here we are dealing with direct managerial actions and not policy decisions, the central government's involvement cannot code as high (code = low).

The Zaibatsu, representing individual families of traditional status and power, emerged as the managing élite, confirmed by the central government as having enormous authority in the areas they managed. Their high coding is thus not invalidated by the fact that the central government prevented the Zaibatsu from forming a politically competitive group by limiting their authority to their assigned sectors of technology (code = high).

Social Control Variable. Even though our concern here is with technology, in the category of social control the central government reigned supreme. It had delegated authority to the Zaibatsu to manage technologies and continued to keep tight reins on them in terms of potential punitive financial or police action should they challenge the government's policy or political primacy. The real instruments of coercion were completely in its hands (code = high).

The Zaibatsu's control over their technological zones of activity was absolute and the imperial government's coercive resources, legal and military, backed their authority (code = high).

Locus of Validation. All of the industrial and military innovations were initiated by the central government in the name of the emperor, the divine head of state. The principle of validation used is clearly "exterior" to any actual person and is rooted in an abstract or symbolic principle (code = high).

The Zaibatsu had their managerial roles legitimized idealogically by their patriotic support of the emperor and so drew their validity from adherence to an abstract principle (rather than as would a founder of an enterprise, for example) (code = high).

The codings for the above data are found in Table VI where they appear alongside the same codings for the late Tokogawa period.

Assessment of the Political Organization: Tokogawa Period

Having assessed the changes in the managerial characteristics displayed by managerial entities in controlling three sets of technological activities, we now turn to the analysis of the organizational characteristics of the Political Zone. As defined in the model, the Political Zone concerns the integration of numerous managerial enterprises into a viable policy. In Japanese society at the time of the Tokogawa Shogunate, these roles were divided among three groups: the Shogun, the Daimyo, and the merchant-bankers. Our purpose is now to derive codes for their organizational status.

An advantage of this model is that the same variables that measure managerial roles organizationally, can be utilized to elicit organization formats of the Political Zone.

THE COMMUNICATIONS COMPONENT

Locus of Authority Variable. The Shogunate operated as a military dictatorship. Politically, if not economically, its control was effective in so far as no competing political adversary was able to successfully challenge its supremacy. It was the most potent locus of political authority in the realm (code = high).

The Daimyo, as feudal lords, especially if they were strong, exercised absolute control over their own territories. Their political power base, both economical and military, rested on the direct exploitation of the peasantry (code = low).

The merchants had little direct political power in a feudal, aristocratic system. Despite their growing economic wealth, they could only apply political pressures through giving or withholding monies to the Daimyo or Shogun (code = low).

Channels of Communications Variable. In organizational terms the Shogunate really represented a powerful feudal clan attempting to operate as a national government. As it had to organize and control large numbers of dispersed feudal entities, it had, of necessity, to create and maintain many channels of communication (code = high).

The Daimyo, though powerful locally, were relatively isolated by the "divide-and-control" tactics of the Shogun. The feudal lords could only enlarge their communication channels via temporary alliances (code = low).

The channels of communication open to the merchants to operate in a political arena could only be devious and indirect in a feudal society based on aristocratic privilege. Their numerous contacts with merchants or even Daimyo only gave them effective commercial influence (code = low).

Complexity of Linkage Variable. The complex political maneuverings of the Shogun required that he knit and sequence many economic and political acts involving scores of clans and families to preserve his precariously held power. Thus his linkages were not only numerous but involved monitoring and adjusting his strategies continually. To accomplish this meant complex dealings with many bureaus, agencies, and hired specialists (code = high).

The Daimyo, even if to a lesser extent than the Shogun, also was the center of a complex set of linkages involving merchants, military retainers, craftsmen, peasants, and petty bureaucrats (code = high).

The merchants, despite the growing complexity of their sphere of commercial action producing new economic linkages, could not be overtly active on the political scene (code = low).

THE DOMAIN COMPONENT

Size-Quantity Variable. The Shogunate and its administrative and social control agencies (inspectors, police, spies, etc.) attempted to control the affairs of the entire country. Thus the units involved were large and integrated (code = high).

The Daimyo, though mirroring this complexity in pattern, did so on a much smaller scale than the Shogunate (code = low).

The merchants, as noted, were controlled by the Daimyo locally and by the Shogun nationally, and hence had no overt official political agencies or tools (code = low).

Heterogeneity Variable. In order to maintain his power, the Shogun depended to a large extent on the cooperation of a broad strata of aristocrats who shared the same values and clan organization. Thus, even as a political entity, the Shogunate was not really differentiated from the general culture pattern of feudal organization (code = low).

The Daimyo, of course, as feudal components of the old Japanese political order, also represent homogeneity rather than specialized differentiation, or heterogeneity (code = low).

Surprisingly, the merchants emerge as a social category tending toward heterogeneity as a result of their widespread commercial relationships unhampered by traditional family or class concerns. Their many-faceted contacts with nobles, craftsmen, peasants, and other merchants was of a different character than that displayed by the Shogun and the Daimyo in this context. Though not apparent on the surface, their form of heterogeneity did have great potential political significance (code = high).

Concentration Variable. Since the Shogunate exercised control over the entire nation, its political activities were indeed widely dispersed (code = high).

The Daimyo's activities, even accounting for alliances with other feudal lords, cannot be coded as approaching the characteristics of the Shogunate (code = low).

The more successful merchants were dispersed throughout the nation and in fact escaped by devious means the nets cast by the feudal authorities to trap and exploit their wealth. In our judgment, for this variable they code higher than the Daimyo, though, obviously still well below the Shogunate (code = high).

THE LEGITIMACY COMPONENT

Boundary of Authority Variable. The Shogun as head of a totalitarian order exercised formal authority over almost all aspects of life, ranging from styles of dress and forms of entertainment to military and political affairs (code = high).

The Daimyo tended to emulate the Shogun but, of course, on a considerably reduced scale (code = low).

Clearly, the merchants had no acknowledged formal areas of political activity to assert any authority, nor did they dare claim any (code = low).

Social Control Variable. The Shogun demanded absolute obedience to his authority and backed his claims by ideological or moral persuasion, but more so by such coercive techniques as internal passport controls, secret police, holding nobles hostage in his palace, and various forms of physical punishment (code = high).

The Daimyo exercised a similar absolute power within their territories and while not attaining the same scale (as measured by a different variable) had the same organizational quality of impact as the Shogunate (code = high).

The merchants can hardly be said to have had any access to formal means of exerting social control (code = low).

Locus of Validation Variable. The Shogun supposedly obtained legitimation of his rule by claiming to be the servant of the emperor. In fact, he ruled on the basis of personal or family power in political and military terms. Despite the rhetoric used, the real locus of validation was not based on abstract principles but on naked force (code = low).

Similarly, the Daimyo, when viewed in organizational terms rather than rhetorical ones, really, despite feudal principle, also based their power on their military abilities (code = low).

The merchants, obviously, must code very low here (code = low).

We have summarized these evaluations of the pre-change state of political organization in Table VII.

Assessment of the Political Organization: Post-Meiji Period

Now we turn to the analysis of the political changes that resulted from the Meiji reforms forty years later. As noted before, in 1910 the critical social categories now have been reduced to two: the central government and the Zaibatsu.

THE COMMUNICATIONS COMPONENT

Locus of Authority Variable. Power was now concentrated in a state bureaucracy ideologically based on allegiance to the divine emperor and was ritually expressed in the new state Shinto religion. Prefectural government replaced the clan as the fundamental local territorial unit. The new central government now made all major policy decisions and acted as a counterbalancing force to prevent predominance of the new industrial-banking Zaibatsu centers of power. The locus of authority in the land was clearly seen as being in the hands of the government in economic, political, or military affairs (code = high).

The Zaibatsu, despite their enormous managerial powers, could only partially participate in basic policy formation. The central government jealously guarded its primacy in such affairs by using every device to warn off contenders. In this case the military and police establishment as an arm of the government ensured its authority. The Zaibatsu had their managerial roles assigned to them (code = low).

Channels of Communications Variable. The new and elaborate system of state and local government which replaced feudal indirect rule by direct controls through linked state/prefectural bureaucracies in chains of command obviously generated many channels of communication (code = high).

Table VII. Coded Assessments for the Political Zone

	Tokogawa Period			*Post-Meiji Period*	
Variable and Component	*Shogunate*	*Daimyo*	*Merchants*	*Central Government*	*Zaibatsu*
COMMUNICATIONS COMPONENT					
Locus of Authority	High	High	Low	High	Low
Channels of Communications	High	Low	Low	High	High
Complexity of Linkage	High	High	Low	High	High
DOMAIN COMPONENT					
Size-Quantity	High	Low	Low	High	High
Heterogeneity	Low	Low	High	Low	Low
Concentration	High	Low	High	High	High
LEGITIMACY COMPONENT					
Boundary of Authority	High	Low	Low	High	High
Social Control	High	High	Low	High	High
Locus of Validation	Low	Low	Low	High	High

Tokogawa Period = Initial state
Post-Meiji Period = Achieved state
Desired state (Western European political organization) codes high in all categories

The Zaibatsu as modern industrialists reflected the organizational complexity that their new technologies implied. The steel or shipping industries are really industrial empires, each with many complex operations which must be integrated by numerous channels of communication. Different industries must be related as they are interdependent upon each other's products and services (code = high).

Complexity of Linkage Variable. The adoption of the European bureaucratic mode of national government and the need to control and plan for the nation as a whole was bound to create a vast and complex linkage pattern (code = high).

Similarly, the Zaibatsu, incorporating the new managerial techniques developed by Europeans and being bound into complex relationships with government agencies, of necessity produced a pattern of complex linkages (code = high).

THE DOMAIN COMPONENT

Size-Quantity Variable. The activities generated by the new technologies in which the government was initially deeply involved required that these be organized on a national and even international scale. The technologies, and the administrative units that directed them, were now numerous and large (code = high).

The Zaibatsu and the activities they managed were locked into an integrated central plan of development. But the new industries being established were by their very nature complex entities as measured in terms of the size-quantity variable (code = high).

Heterogeneity Variable. Even though in 1910 Japan was a complex nation-state with a rich and diversified social organization, the political power strata represented a narrow and homogeneous strata of favored aristocratic families. The new pseudo-democratic official institutions such as the Diet remained without effective power (code = low).

The Zaibatsu, therefore, also code low because the membership in the new oligarchy was drawn from the same narrow, homogeneous group as was the state bureaucracy (code = low).

Concentration Variable. The central government was organized on an empire-wide basis and so falls well into the "dispersed" category of this variable (code = high).

The Zaibatsu, for reasons already well exposed, managed industrial operations well dispersed in space (code = high).

Boundary of Authority Variable. The new regime was almost as massively restrictive in its control of mundane, as well as high, state affairs as the former Shogunate. The Meiji reformers introduced Shinto as a means of asserting total authority over the population. As a well-defined entity, the central government emerged as the unambiguous ruler of Japan (code = high).

The Zaibatsu, by participating in the great patriotic venture of modernizing Japan and as instruments of achieving national goals of political autonomy and industrial-military power, were able to assert their delegated authority over many areas of economic and industrial activity (code = high).

Social Control Variable. All the instrumentalities of the Shogunate, punative laws, etc., were now made much more efficient and sophisticated. The totalitarian character of the state was undiminished (code = high).

The Zaibatsu cooperated with the central government in maintaining absolute state control, in part to ensure their position as a privileged group (code = high).

Locus of Validation Variable. The central government and the emperor were now joined as a unit of validity. Such a claim to power is based on an abstract principle—the divinity of the emperor (code = high).

The Zaibatsu legitimated their authority, likewise, by their obedience to the emperor (code = high).

Codings for the Post-Meiji political organization are given in Table VII.

Analytical Discussion

Having elicited coded values for empirical technological and social organizational behaviors in the Japanese case, we can proceed to analyze the process of transformation of system-states. The degree to which the achieved organizational state of Japanese society met the stated requirements of a desired condition sought by political leaders can now be measured and the policies initiating the transformation judged as to their degree of success or failure.

To apply the model, the value loadings of components that define organizational system-states are derived from averaging the measurements of the set of variables that compose the components. Then these components are examined to determine into which of eight possible analytical patterns they fall (Table II). The process of transformation is mirrored in changes occurring in the value loadings in components when organizational states are changed. For example, in Table III to change Organizational State G (low, high, high) to State D (high, high, low) requires that the averages for both the Communications Component and the Legitimacy Component be reversed. Since component values are averages of variables, at least two of the variables for each component will have to undergo change to effect this change (Table IV).

The Japanese case is universally acknowledged as an example of successful transformation from a pre-industrial to a modern industrialized state. Upon examination of the results of our translation of the real events into the variables of our model, we find that the technological aspects of the transformation have indeed followed the rules of successful transformation. However, the corresponding managerial and political organizational structures exhibit certain failures to achieve an organizational status matching their ideal desired stated aims. In all three cases the change in technological states is from a condition of low, low, low to one of high, high, high—*i.e.,* from type E in Table II to type C. Upon consulting the values for the desired state, we find that Western European technology, the model being followed, also codes as Type C. Therefore, we can categorically state that the Japanese were successful in their desired end; they achieved the desired state in their technological program.

Examination of Table VIII shows that this transition from Organizational State E to C involves a change in the values for all three compo-

Table VIII. Possible Organizational States for Components of a Zone

	Analytic Pattern (value loadings assigned)		
Organizational State	*Communications Component*	*Domain Component*	*Legitimacy Component*
A	High	Low	Low
B	High	High	Low
C	High	High	High
D	High	Low	High
E	Low	Low	Low
F	Low	Low	High
G	Low	High	High
H	Low	High	Low

nents. Our analysis of Japanese technology (as represented by our three selected areas) shows that all three components were changed by the appropriate manipulation of the constituent variables in the correct direction—from low to high. We may then conclude that with respect to technology, the Meiji reformers followed the correct and necessary strategy in making these adjustments.

However, the introduction of novel technological features of organization is only one of the core issues to which we address ourselves. Our concern is with the interface of the technological *and* the social organizational realms. In this area we find that Tokogawa management was concentrated in the hands of the Daimyo and the associated merchants who were closely tied to them. We note that managerially the organizational pattern of the Daimyo is characterized by a component value of high, low, high while that of merchants is low, low, low. The Daimyo constituted a real managerial class, who, in practical terms, had more than enough capacity to deal with the Tokogawa technology, but our model indicates real weaknesses in the Domain Component. Confronted with this state of affairs, the Meiji reformers merged the Daimyo nobles and merchant-banker categories into a single economic and industrial entity—the Zaibatsu. This tactic capitalized upon the strengths of the Daimyo in the Communications and Legitimacy Components as well as the merchants' high values in the Concentration variable of the Domain Component, thereby making it possible, by raising the value of a single variable (the Size-Quantity variable), to raise the entire Domain Component. This was truly a master stroke, in that it created the high averages for all three components needed to introduce and develop the new technologies. By 1910 the Zaibatsu had a well-matured managerial grasp of the new technologies.

Unlike the European situation at that period, in late nineteenth century Japan the central government did play a powerful and direct managerial role in industrial development. But here the model displays certain weaknesses in its performances in a direct managerial capacity. The desired state for any managerial entity required high loadings for all of the variables. It is true that the averages of the variables in all three components attain a high loading permitting the technological development to be successful. But the degree of success is not absolute because the Meiji reformers fell short of raising the values of all of the variables. This ability of the model to expose the latent weaknesses in organizational growth is of great importance in policy analysis. By retaining many feudal features in their managerial organization the Japanese were restricting themselves to a more limited organizational base than that developed by certain Western powers.

The weakness of the central government's managerial role, since it falls on the variables—Locus of Authority, Size-Quantity, and Boundary of Authority—has certain consequences. Historically, the imperial government, and not the Zaibatsu, was the initiator of technological change, and the three weak variables are the very ones that are directly concerned with managerial innovation. In contrast, for these variables, the Zaibatsu display relatively greater strength in the initial phase of change. But since the Zaibatsu retain older, powerful conservative elements which still exhibit feudal features, their primary control over technological activities in the last phases of change results in drags on managerial development which the government cannot effectively counter. The government indeed can control the policy and even fundamental economic aspects of technological change, but it is no longer a direct managerial agency by 1910, and could not make the conservative, controlling Zaibatsu maintain a progressive attitude toward technological change. Managerially speaking, Japan leapt technologically into the late nineteenth century and then stayed there until very recent times (post World War II).

A similar series of weaknesses emerges from the analysis of the Political Zone of organization. Here the glaring weakness in both the central government and Zaibatsu circles is displayed as an excessive homogeneity—as measured by the Heterogeneity Variable.

Again, the components of political organization all average high so that the innovative developments go forward, at first. The tendency toward homogeneity displayed reinforces the conservative noninnovative behavior already exposed in the Managerial Zone affecting technology. Japan, in 1910, did not exhibit the structural capacity to organize the continually changing technical and managerial innovations that characterize the current, fully industrialized state.

The social elements that made up the new Post-Meiji regime, in fact, still represented the ancient aristocratic and traditional merchant-banking interests. No new social classes really emerged from the industrial reforms to assume a political role. The ruling strata, while appearing radical and innovative at the technological level, was actually intensely conservative and narrow in outlook. The variable that reflects this best is the Heterogeneity variable which codes low for both the central government and the Zaibatsu.

In the model the required transformation would change the Tokogawa Organizational State B (high, high, low) to State C (high, high,

high). This was achieved on the component level, but with the specific weaknesses displayed in certain variables. The full desired state was not achieved.

We believe that our form of analysis is more precise in specifying the particular empirical situations and their consequences than other historical and sociological analyses of the same materials. The apparent immediate success of the Japanese program of modernization tended to blind both the Japanese and foreign observers to the organizational weaknesses latent in the chosen strategies. In the end, policy evaluation must focus upon defining in operational terms what is meant by a proposed strategy of change. We believe our model is a step forward in this direction, because it clearly illustrates the dynamic interaction between technology and social organization in a single frame of reference.

6 Case Study 3: Organizational Collapse in the Face of Radical Change: Manchu Dynasty of China

Introduction

The content of this chapter concerns the tragic failure of traditional Chinese society in the last period of dynastic rule to master European technology.[1] However, it is not our intention to contribute new information on this sad chapter in Chinese history but to utilize this well-documented case to illustrate the properties of our model which relates technology to social organization in a new manner. It is precisely because of the patent failure of the Chinese political system to

[1]The data utilized in this analysis were derived from the following sources: Beckman (1962), Kiernan (1970), Lowe (1966), and Ssu-Yii and Fairbank (1954).

accommodate itself to change, for reasons well known to any historian of the empire (see Bibliography), that it was chosen. We indeed know why the Chinese political system failed. It refused to accept new social and political institutions and clung to outmoded bureaucratic and balance-of-power strategies that time and time again failed to meet the challenge of European and Japanese economic and military imperialism. Rather than give up its ideologies and deeply venerated Confucian social order, it preferred, literally, to die. It could not invent new technologies to counter those of Europe, nor could it even adopt or imitate the technical devices such as railroads, steamships, crew-served weapon systems, etc., necessary to defend its interests because it would not permit a new class of technicians and scientific or professional managers to arise. It demanded that its exquisitely civilized, but scientifically illiterate, civil servants achieve mastery of highly technical enterprises—a hopeless assignment for them. The traditional merchants and monied entrepreneurs would not invest their capital in industrial or other productive enterprises under conditions of grave instability. Still other contributing causes can easily be noted, but these suffice.

We, too, arrive at and confirm such conclusions, but do so in a novel way. It is our contention that all the above circumstances are empirical expression of a basic failure of organizational patterning at the political level to design effective organizational formats at the managerial level and thereby, in turn, permit the organizational conditions characteristic of advanced nineteenth century Western technology to be adopted. In the previous chapter the Japanese case study was used to expose a basic axiom of the model, that the organizational properties of a social entity must have the requisite (measurable) amount of organization to match, at least, that expressed in the organizational format of the technology it seeks to manage. The Japanese management innovations of the late nineteenth century were presented as an example of how successfully to incorporate new technologies following the principles suggested by the model.

The same principles hold for the development of *strategic policies* by a political body which is meant to encourage and foster technological change. The strategy of change adopted by a policy-making body must provide its managing agencies (those bodies that actually carry out technological tasks) with the organizational properties that at least match the level of complexity that the technologies display. It is the ability of our model to *measure* the complexity of organizational states and match the social organizational components against technological activities that makes evident the degree of success or failure of a policy. That the Chinese failed in their policies is obvious, but to determine by how much of a margin and in which specific aspects of organization the failure most accrued, and what other possible strategies than the ones they used might have been attempted—these we believe to be novel aspects of analysis.

The model, because it is capable of projecting a series of different possible strategies of change from a given initial, historical point of departure, makes it possible, thereby, to consider hypothetical "futures". What happened to the Chinese is a case of their pursuing one of several scenarios of change potentially open to them in the late nineteenth century. The strategy of change they chose violates the principles of successful organizational change as posited by the model. (The Japanese, note, did succeed in the same time period.)

This presentation focuses on the critical problem now so commonly displayed by various developing countries in the twentieth century—how to make the proper strategic choices from among an array of possibilities in planning to adopt new technologies. The model offers a tool for analyzing various courses of action open to any policy given its present state of managerial and technological organization and suggests procedures to ensure that the necessary organizational properties of its institutions come about. The empirical example of China shows, we believe, the fatal consequences of choosing, willfully or otherwise, the wrong strategy of change.

The first step is to formulate a model of the organizational properties of Chinese technology as it existed in 1860 in order that this can be compared with the state of Chinese technology in 1910, the last year of the Manchu Dynasty's reign. It is also necessary to determine what state of technology the Imperial Chinese government wished to achieve. In this case it is clear that the technologies of Western Europe were the models. Hence, a coding of the specific technologies as they were found in the last half of the nineteenth century in the advanced industrial nations provides the data for a comparison of the existing technology of China with a desired state. The actual, initial state of organization (once it is determined by coding descriptive data from that period) is located in Table IX as one of eight possible organizational formats; the desired state of organization is then also located in the table and noted to be one of the eight ideal types. This table then shows which variables and components

Table IX. Possible Organizational States and Their Transformations

	Analytic Patterns (Value Loadings Assigned)		
Organizational States	*Communications Component*	*Domain Component*	*Legitimacy Component*
A	High	Low	Low
B	High	High	Low
C	High	High	High
D	High	Low	High
E	Low	Low	Low
F	Low	Low	High
G	Low	High	High
H	Low	High	Low

must be changed in order that the desired organizational state might be achieved.

Actually, as in the Japanese case, only three critical technologies were chosen to represent the zone in the model: transportation and communications, the military, and heavy industry, specifically steel manufacture. These three technologies constituted the heart of the Chinese policy which sought through acquiring new technologies to defend itself against Western European, Russian, and Japanese imperialism because, to the end, the Imperial Throne and its closest counsellors refused to believe its ancient social institutions and culture were inferior to those of the barbarous foreigners and therefore should be abandoned. Some modifications might be tolerated, but nothing radical would receive serious consideration.

The time period of 1860 was chosen as the base from which to measure change because it represents the last era when the Manchu Dynasty could still claim to rule over the entire empire. The Tai-Ping Rebellion was being finally contained, though at hideous cost, and the various encroachments of foreign powers had not yet disintegrated the political hold of the throne over interior China. The vast civil service bureaucracy united China politically and the moral authority of the emperor still preserved some of its potency.

NOTE ON CODING PROCEDURES

The two zones comprising technological and managerial aspects of organization will be presented as condensed descriptive summaries of conditions contrasting the 1860 and 1910 periods. The data will not be coded variable by variable as this procedure would be much too lengthy. Only the final value loadings for the components of these two zones will be given at the end of each descriptive section. However, the Political Zone, the main focus of this exposition, will receive a full analytic treatment in which each of the nine variables will appear in order to illustrate how values were assigned to political organization. We fully expect that not every reader will accept our coded values for the descriptive materials offered, but we ask indulgence on the basis of a recognition that it is really the model in its entirety as a tool for techno-social change analysis that should be judged as to its value, and not every specific coded item.

Assessment of the Initial Technological State: 1860

TRANSPORTATION AND COMMUNICATIONS

Technologically, Manchu China displayed most of the characteristics of pre-industrial societies despite its unusually complex political development. There was a preponderance of human- and animal-powered vehicles for land transportation and communications. The Grand Canal, which connected the Yellow and Yangtse rivers, functioned as a vital factor in the integration of the vast territorial expanse within the empire. The canal network provided a cheap and easy form of transportation of goods and men despite the fact that no maintained road system existed, but while they did possess elements of a viable internal transportation system, it was really confined to certain inland routes. Though the empire had naval craft capable of oceanic voyages and the central government maintained a post-rider system for its important diplomatic correspondence, areas of inland China isolated from the river-canal network could only be reached by using the slow caravan routes. By 1860, the effects of massive rebellions and widespread local banditry had forced the government to abandon, ever increasingly, its efforts to maintain the vital canal system, resulting in a drastic weakening of the traditional transportation-communication structure.

MILITARY TECHNOLOGY

The military technology of Manchu China was woefully primitive in comparison with that available to Western European powers. Peking could only command the remnants of an effective standing military force—the famous "banner" armies were now lowly regarded Manchu cavalry garrisons, located in the provinces, quite unable to face any determined or disciplined foe. The central government possessed a local military force at Peking, the Peking Field Force, which, while armed with modern rifles and some artillery, was small in size and really represented a large palace guard. The provincial bureaucracy could attempt to marshall local militia, but these were poorly armed and hardly a match for its foreign adversaries (as was demonstrated in the Boxer Rebellion, later). Some European military aid had been utilized in suppressing the Tai-Ping Rebellion, but such staff innovations were mercenary in nature and never were effectively incorporated into the command structure.

There was no heavy industry to match European standards of technology. Productive enterprises such as silk or porcelain manufacture were officially promoted by the state, but these activities created trade goods for the domestic and foreign markets. Despite the technical and aesthetic qualities of Chinese artifacts, their production represents the level of cottage industry. Most of the goods were used in the locales where they were manufactured.

Assessment of the Desired Technological State

The desired state of technological development, at least for the reform-minded ministers within the dynasty, was that of contemporary Western society. In the latter half of the nineteenth century, the Western powers commanded vast technological resources. Steam-powered transportation and telegraphic systems made for effective control over even the distant territories under their spheres of influence. A heavy steel industry assured abundant supplies of the critical metal necessary for the manufacture of railway engines, steamships, and bridges. Militarily, the new railway and telegraphy systems, the accurate rapid-fire weapons and field artillery, the armored and speedy warships, and the trained technical and professional military staffs all made European powers relatively invincible when in conflict with pre-industrial military forces.

This desired state codes on the component level as *high, high, high.*

Technological Assessment of China: 1910

The various attempts to introduce the above complex technologies into China cannot be treated in detail here. But a summary conclusion is easily made. In only a few isolated instances were any major technological innovations successfully incorporated into the fabric of Chinese technological organization. In nearly every instance such attempts ended in failure, and the net effect was that Chinese technology retreated to a lower level of effective organization than pertained in 1860. Not only were the new technologies not mastered, but the central government and provincial bureaucracies failed to maintain the intricate canal and irrigation systems that constituted the heart of the transportation and communications facilities of the old techno-social order.

Some attempts at introducing railroad and telegraph lines were made and a steamship construction yard was established at Tiensien. However, these were later purchased by the central government in one of its retrograde spasms and dismantled. Other rail and mining enterprises were established by foreign concessions. Those that were financed and managed by European, Japanese, or American interests operated in entirely autonomous foreign enclaves under extra-territorial privileges. No steel industry was established, nor, indeed, was there any technological enterprise of any complexity involving transportation or communication effectively in Chinese hands by 1910.

In the area of military technology, all attempts of the Chinese to procure and utilize modern armaments ended in disaster. A Chinese fleet was built but was destroyed by a superior Japanese naval force in 1894. Western arms suppliers limited their sales to old, obsolete weapons, since these were the very Western powers who were seeking to establish spheres of influence, if not outright possession of territory, in China. They did not wish to strengthen militarily a China they sought to dominate.

The coded assessments for Chinese technology in 1910 thus remains *low, low, low* (see Table X for a summary of the two time periods).

Assessment of Management of Technology: 1860

The model does not require the consideration of the entire social organization of China, only the identification of those agents and agencies that actually controlled the three technological activities being examined. In the case of the Late Manchu Dynasty, only three categories need be considered as *managerial entities:* the central bureaucracy, located in Peking; the local provincial bureaucracies; and the influential merchant families, who represented commercial and capital interests. The rich and prestigious rural gentry, though obviously a dominant social and economic component of the social order, nevertheless, need only concern us as members of the civil service bureaucracy.

The central government, managerially considered, was a complex political entity which did not carry out, directly, any managerial task interfacing with an actual technological activity. The imperial throne and its massive bureaucracy directed local bureaucratic, administrative agencies to initiate and carry out specific projects affecting arms production or procurement, maintenance of transportation and communication routes, and keeping order. In terms of priorities, keeping order and

Table X. Assessments for Technology of Imperial China: Initial and Achieved States

Variable and Component	*Transportation and Communications*		*Military Technology*		*Heavy Industry*	
	1860	*1910*	*1860*	*1910*	*1860*	*1910*
MAN-ARTIFACT COMPONENT						
Locus of Dominance	Low	Low	Low	Low	Low	Low
Complexity of Assemblage	Low	Low	Low	Low	Low	Low
Phases of Unit Interaction	Low	Low	Low	Low	Low	Low
TASK COMPONENT						
Serial Characteristic	Low	Low	Low	Low	Low	Low
Operations-Output Relations	Low	Low	Low	Low	Low	Low
Output Form	Low	Low	Low	Low	Low	Low
TASK COMPONENT						
Setting Structure	Low	Low	Low	Low	Low	Low
Locus of Input	Low	Low	Low	Low	Low	Low
Autonomy	Low	Low	Low	Low	Low	Low

1860 is the initial state; 1910 is the achieved state. The Western European desired state codes high for all variables.

collecting taxes were paramount. This administrative structure, though badly strained by violent internal disorders and foreign incursions, nevertheless, still maintained organizational cohesion.

The primary managerial responsibilities, as it concerns our three areas of technology, was officially assigned to the provincial civil servant category, especially the Mandarins. Until the nineteenth century the provincial bureaucracy was able to fulfill its assigned managerial roles because these roles were largely administrative in nature. The relatively simple pre-industrial technology could safely be left in the hands of the traditional entrepreneurial elements (*i.e.*, the merchants and artisans) and manpower for the more massive and time-consuming tasks of maintaining the canals could be carried out by peasant conscript labor supplied by cooperative village elders, who accepted this traditional obligation.

As the nineteenth century progressed, the impact of foreign technological developments in communication, transportation, heavy industry, and weaponry led the imperial government to assign tasks to its provincial agents in which they were now officially responsible for developing and incorporating these new technologies in order to strengthen the empire. Unfortunately, these provincial bureaucrats could no longer use the traditional productive format to achieve the new technological goals. The traditional pattern being highly individualistic, with each social category (gentry, merchants, rich peasants, artisans, etc.) jealously motivated by self-interest, proved highly unsuitable to meet the organizational conditions required by such new technologies as the railroad or steel production. This is the key to our analysis, since it is an important hypothesis, derived from the model, that the social organizational constraints exhibited by a managing entity for operating a given technology must meet the constraint prerequisites for that technology.

An examination of Table XI shows that in 1860 the central government had effective resources mainly in its "legitimatizing" role for maintaining the old technologies. The provincial governments also displayed certain positive resources mirrored in high codings for the heterogeneity and validation variables. For the merchants, their managerial potential is displayed in high coding for the linkage and concentration variables. However, while there was ample organizational complexity to cope with the traditional technology in the old order, the ways in which the favorable components of organizational format found in the three groups were used in concert or held separated with reference to the new technologies is an entirely different matter. At least in 1860, the provincial bureaucracy could perform its managerial tasks in terms of its tradi-

Table XI. Managerial Organization of the Manchu Dynasty

	Central Government		*Provincial Bureaucracy*		*Merchants*	
Variable and Component	*1860*	*1910*	*1860*	*1910*	*1860*	*1910*
COMMUNICATIONS COMPONENT						
Locus of Authority	Low	Low	Low	Low	Low	Low
Channels of Communications	Low	Low	Low	Low	Low	Low
Complexity of Linkage	Low	Low	High	Low	High	Low
DOMAIN COMPONENT						
Size-Quantity	Low	Low	Low	Low	Low	Low
Heterogeneity	Low	Low	High	Low	Low	Low
Concentration	Low	Low	Low	Low	High	Low
LEGITIMACY COMPONENT						
Boundary of Authority	Low	Low	Low	Low	Low	Low
Social Control	Low	Low	Low	Low	Low	Low
Locus of Validation	High	Low	High	Low	Low	Low

1860 is the initial state; 1910 is the achieved state. The Western European model for the desired state codes high in all variables.

tional format, *if* disorder, rebellion, or communication breakdown did not interfere. The merchants, likewise, were quite able to carry out their middleman function of linking the numerous producers and consumers *if* social conditions were stabilized.

By the early twentieth century, the full impact of the processes initiated in the nineteenth century to foster technological innovation had exhausted the administrative capacity of the provincial bureaucracy and disrupted the operations of the merchants through a rapid deterioration of local social and economic conditions, epitomized by the emergence of warlords as the politically dominant provincial powers. In effect, managerial activities as programmed by the central government ceased not only for new technologies but for the traditional ones as well (*e.g.*, the canal and irrigation works). Those technological activities that were functioning tended to be foreign-owned and controlled, such as shipping, mining, transportation, and communication facilities. The uniformly low coded values for 1910 technology shown in Table X illustrates the collapse of the Manchu Dynasty's managerial capacity.

Assessment of the Initial Political State: 1860

Having completed a highly condensed analysis of technological and managerial aspects of modelling the late Manchu organizational structure, we will now turn to a more extensive treatment of the format of political organization, the central concern of this presentation. The more detailed procedures of exposition requires presentation of data which are coded by each variable, displaying the organizational characteristics of the two politically significant components of political power, the central government and the provincial bureaucracy.

THE COMMUNICATIONS COMPONENT

Locus of Authority Variable. The imperial bureaucracy was centered on the person of the emperor (later, empress) and governmental tasks were delegated to various ministries (code = high).

The provincial bureaucracy, though in theory representing the authority of the emperor, in fact was in an ambiguous position. Officials were dependent on the imperial regime for authority and upon the goodwill of local influential elites, the rich gentry, merchants, and even village heads, to carry out administrative assignments (code = low).

Channels of Communications Variable. Within the political structure of the imperial bureaucracy, the numbers of channels of communication were extremely numerous, and actually, at times, designed to block action through a subtle set of checks and balances between and among its agents and agencies. As a foreign dynasty, the Manchus were in the dangerous position of being dependent on a mainly Chinese bureaucracy and feared power accumulations in their hands. But as the variable measures the *quantity* and not the quality of the communication channels, a high value is assigned (code = high).

The provincial bureaucracy, as part of the imperial system of administration, had to develop its own local contacts with every strata of society. It therefore exhibited large numbers of channels of communication (code = high).

Complexity of Linkage Variable. The imperial bureaucracy had highly complex linkage patterns characteristic of bureaucratic order that controlled the affairs of a vast realm (code = high).

The provincial bureaucracy was part of this imperial system (code = high).

THE DOMAIN COMPONENT

Size-Quantity Variable. As China was a vast empire, the number of activities to be organized politically were both large and numerous, for example, to integrate and manage the various provincial administrative organizations (code = high).

The provincial bureaucracy repeated this pattern at the local level, but the local level here concerned huge provinces, often larger than European nations (code = high).

Heterogeneity Variable. At the level of imperial government, a narrow and homogeneous group of Manchu and Chinese aristocrats and powerful families allocated to themselves supreme power. The varied and complex Chinese social orders did not participate in decision-making processes (code = low).

The provincial bureaucracy, being part of the same system, represented the same interests, even though their appointments were attained through the rigorous civil service examination procedures. Even those recruits to the system representing families of low social status rising to positions of authority, now exhibited the same values and interests as the conservative, exclusivistic literati strata (code = low).

Concentration Variable. The imperial government was highly concentrated in its activities, centering its activities in Peking where the court jealously hung on to centralized authority. Only nominal direct powers in fact could be exercised over the provincial authorities, who, in the end, were often left to their own devices. The throne exercised such indirect controls over its dispersed administrators as threats of removal from office, or issuing urgent (often conflicting) directives and admonitory edicts (code = low).

The provincial bureaucracy, represented by the Mandarin in the provincial capital and the lesser officials in outlying districts, was well dispersed everywhere (code = high).

THE LEGITIMACY COMPONENT

Boundary of Authority Variable. In theory, the emperor had complete and total authority over every aspect of life, guided by Confucian principles (code = high).

The provincial administrators obviously shared the same massive authority (code = high).

Social Control Variable. With the decline in effectiveness of the Manchu cavalry garrisons (the "banners"), the imperial government became entirely dependent upon local, provincial authorities to enforce, as best they could, its decisions and carry out police or military operations (code = low).

The provincial governments were now almost totally in charge of operations that controlled local populations. However, their dependence on local militia and peasant drafts weakened their actual punative capacities (code = low).

Locus of Validation Variable. The imperial government epitomized authority which rested on an abstract principle, the "mandate of Heaven" concept, and the Confucian ethic that spelled out the norms of all social and political behaviors (code = high).

Assessment of the Political Organization: 1910

The year 1910 represents the last year of the existence of the Manchu Dynasty and provides a convenient date for the assessment of fifty years of attempts to accommodate to the stresses of meeting the technological and political challenges.

Locus of Authority Variable. The attempts by the reform elements in the imperial bureaucracy to introduce Western technology were systematically weakened or sabotaged by the traditionalists in the government. There was no real power base within the court circle to which the reformers could attach themselves. The primary aim of the imperial court and great families associated with the powerful administrative bureaus was to maintain the status quo. The leadership efforts of the progressive wing of literati (including the emperor himself at one time) were crippled by their isolation from the political instrumentalities necessary to maintain a policy of modernization (code = low).

The provincial bureaucracy, even when headed by reform-minded bureaucrats, was without effective authority because of the imperial government's inability to formulate and maintain firm policy decisions. The local officials were forced to assume greater and greater roles of a policy type with an ever declining resource base in tax money and military authority. They were forced to make alliances with local warlords or become their puppets (code = low).

Channels of Communications Variable. The imperial government made numerous attempts to introduce technological changes such as establishing railroads, constructing a naval yard, creating a telegraphic system, etc. However, the mode of implementation was through an existing bureaucratic organization whose instincts and loyalties led them to be ambivalent or actively hostile to such changes which threatened their social position. It was really beyond their real capacity to perform the technically specialized tasks, given their training and ideological commitments as literati. The net result was that none of these attempts succeeded. In other words, the numerous old channels of communication were inappropriate and no new ones adequate in format to the new technologies were developed (code = low).

The provincial bureaucracy displayed a slight increase in local channels of communication involving some modern technological activities, such as acquiring steamship transport or training and equipping armed forces, but as these very activities now depended upon foreign technical and professional personnel and sources of supply, the net effect was to reduce the control of the Mandarins over such instrumentalities (code = low).

Complexity of Linkage Variable. The imperial government made several attempts to promote modern technologies through the traditional linkage structure of the bureaucratic system. As already pointed out, this ancient structure incorporated an inherent strategy of system-maintenance through checks and balances and was totally ill-suited to unify the various departments and agencies in following a consistent strategy of radical innovation. The fundamental ambivalence within the imperial court toward technological innovation and even greater antipathy toward basic social and cultural institutional change precluded using governmental resources to implement a genuine strategy of change. The linkage mechanisms within the government sabotaged rather than fostered change (code = low).

The above conditions pertained at the provincial level. The bureaucratic apparatus hindered rather than assisted change. The continual stream of diplomatic dispatches from Peking demanding that they carry out complex technological and military tasks requiring radical innovations beyond their organizational and personal technical capacities only led to a decline in the effectiveness of the existing and already inappropriate linkage mechanisms. The new technological tasks required far more constraints because of their greater need to be orchestrated in disciplined, phased sequences (*e.g.,* managing a railroad). The political task was how to link many such operations—rail, steel, bridge, mine—into line with political policy (code = low).

THE DOMAIN COMPONENT

Size-Quantity Variable. There was a real decrease in the size and numbers of organized activities that the imperial government controlled as compared to 1860. The productive elements within the provinces, the merchants and artisans, increasingly diminished their functional interface with the central government and retreated into localism. The government in Peking thus lost contact with the various components of the political system. The number of units in fact got smaller in size and quantity. With the country effectively in the hands of unfettered warlords and rapacious foreign enclaves enjoying extraterritorial spheres of economic influence, the imperial court no longer ruled the country (code = low).

The above conditions held for the provincial level as well (code = low).

Heterogeneity Variable. All attempts to introduce heterogeneity in the social composition of the political bureaucracy failed as the traditionalists refused to accommodate to the changes suggested by the reformers. These reformers had proposed a number of political institutional changes which would have brought nontraditional elements of the population into the government via an elected assembly: the Westernization of the education system could break the monopoly of the literati class, etc. Although such proposed innovations initially gained imperial backing, these measures failed to achieve actual implementation (code = low).

Since no reforms were instituted, the provincial bureaucracy continued to be recruited in the traditional manner. In fact, throughout large areas of China, personal rule by warlords had substituted for officially appointed authority (code = low).

Concentration Variable. As the imperial government lost administrative control of the provinces, its ability to govern was more and more localized to the Peking Court itself (code = low).

In the provinces, where the official bureaucracy had any semblance of authority, it had shrunk back, similarly, to the immediate locales of the administrative capitals (code = low).

THE LEGITIMACY COMPONENT

Boundary of Authority Variable. There was a progressive decline in areas of jurisdiction of the imperial government both through surrender of sovereignty to the foreign powers and through loss of moral authority over areas still nominally ruled by the dynasty. The progressive failure of the throne to protect the Chinese from foreign encroachments or to maintain internal stability led to the Manchu Dynasty's loss of its coveted mandate of Heaven source of authority (code = low).

As representatives of the throne, the provincial officials suffered the same fate as the imperial court (code = low).

Social Control Variable. The imperial government obviously failed to introduce modern weapon systems and command facilities necessary to reestablish an effective imperial army through its own resources. It thereby was victimized by a dependence on foreign sources of arms and mercenary leadership, which resources were equally available to competing Chinese dissident elements, namely, the now powerful warlords. They could not mobilize, both technologically and administratively, to coerce rebellious elements within the country or to resist foreign incursions (code = low).

At the provincial level, the Mandarin-headed bureaucracy could no longer mobilize a local constabulary to compete with the bandits and warlords (code = low).

Locus of Validation Variable. The loss of the mandate of Heaven

Table XII. Coded Assessments for Political Organization in Manchu China

Variable and Component	*Central Government*		*Provincial Bureaucracy*	
	1860	*1910*	*1860*	*1910*
COMMUNICATIONS COMPONENT				
Locus of Authority	High	Low	Low	Low
Channels of Communications	High	Low	Low	Low
Complexity of Linkage	High	Low	High	Low
DOMAIN COMPONENT				
Size-Quantity	High	Low	High	Low
Heterogeneity	Low	Low	Low	Low
Concentration	Low	Low	High	Low
LEGITIMACY COMPONENT				
Boundary of Authority	High	Low	High	Low
Social Control	Low	Low	Low	Low
Locus of Validation	High	Low	High	Low

1860 represents the initial state; 1910 is the achieved state. The Western European model for the desired state codes high in all variables.

by the Manchu Dynasty effectively removed any viable source of legitimacy to their rule based on an abstract principle (code = low).

The provincial bureaucracy thereby also lost its source of validity (code = low).

The above analysis of political organization is summarized in Table XII.

Analytic Discussion and Conclusions

The aim of this chapter was to show how the model can analyze the critical interfaces between technological, managerial, and political institutions involved in massive technological change. By translating all the myriads of empirical data about economic, political, and technological conditions into their zones of organization and giving each zone an overall value, it is possible to compress the data into a synoptic organizational picture of a whole society undergoing change. This total zonal value is attained by averaging the values of the three constituent components of each zone. The change of a given society's organizational format into another, will fall into one of the eight conditions displayed in Table IX. The value of this analysis for policy making is that by inspection of the table, certain future organizational alternatives of interfacing technological, managerial, and political behaviors should be evaluated as more desirable than others. The Manchu Dynasty, therefore, in 1860 can be described synoptically as follows: Technological Zone—low, Managerial Zone—low, and Political Zone—high. This represents System State F in Table IX. Because the Manchu Dynasty sought as a primary objective to preserve its traditional political organization it had to maintain high loading in the components of the Political Zone. Therefore, the model suggests there were three possible alternative (future) states open to them under conditions of change permitted to occur in the other two zones. Since the dynasty was under drastic political and economic stress at the time, the maintenance of the total existing state of organization (in all three zones) was not a viable choice of strategy. The three possible choices of system-state format indicated by the model are: Organizational State C (high, high, high); Organizational State D (high, low, high); and Organizational State G (low, high, high).

These three states constitute the only scenarios for strategies of change that would preserve the empire with the dynasty still firmly in control. The first scenario, that is, movement from State F to State C, required that the organizational format of European managerial and technological organization be adopted. The Manchu Dynasty did not consider this to be an acceptable strategy. Such a strategy implied placing strong managerial autonomy in the hands of either provincial bureaucrats, professional administrators, or foreign experts and all such moves threatened the political monopoly of the imperial court and traditional literati elites. The Manchus, as a foreign dynasty, especially feared placing any new locus of power, via technology, in the hands of subject Chinese elements or foreign governments.

The second scenario, movement to Organizational State D (high, low, high), would have required the implementation of technological innovations without resorting to creating a separate managerial entity outside the traditional bureaucratic order. This, indeed, was what the imperial court appeared to favor as a strategy. This, however, proved impractical as a basic strategy since the highly specialized and complex requirements of modern industrial, military, and communication technologies demanded that direct managerial controls be exerted by experts trained to high technical standards. The requisite organizational characteristics were not available to the traditionally trained bureaucrats, as we have seen.

The third scenario, movement to Organizational State G (low, high, high), though logically possible, was clearly a hopeless choice because it would actually ensure that no technological change would come about. The only difference between this alternative organizational format and the actual conditions of 1860 would be an increase in the organizational values of the Managerial Zone without a change in the Technological order. This state would be the obverse of what the Manchu Dynasty consciously sought, which was technological change without allowing new managerial entities to arise to compete with it.

In effect, there were only two viable scenarios to be considered: movement to State C or to State D. The former strategic alternative required the fostering of local and more autonomous managerial capacities, perhaps by a new professional, managerial class. The latter could have been accomplished only through the utilization of foreign management personnel. Clearly, the dynasty was conflicted between the choice of one of these two alternatives, both of which were considered distasteful, or of remaining in its untenable, rapidly deteriorating situation. The inability of the throne to settle firmly on one of these possible alternatives led to a series of partial commitments to each in turn (*e.g.,* use of foreign administrative staffs in some of its technical agencies). Such policy decisions were inevitably and quixotically reversed, due to misgivings as to the possible effects on the centralized power structure in Peking. The net result was lack of *any* consistently supported strategy being followed. The emperor (and empress) and the hard-core traditionalists exhausted the dynasty's remaining prestige in futile efforts to maintain itself in power by ideological appeals to preserve the ancient Confucian order rather than by instituting realistic changes in basic political and technological institutions.

A major conclusion emerges, namely, that the political function in periods of imminent changes must define and consistently apply a strategic policy correctly chosen to maximize the principles of adequate organizational responses. The failure to do so cripples *managerial* efforts (even if these are well designed) to interface with technological behaviors. A second conclusion is that even a correct strategy of change (as defined by the model) when selected by a political entity can be ineffective unless the managerial capacities of the social entities chosen to effect policy decisions are designed with sufficient internal organizational constraint to match that required by the technological tasks they control; and further, that they be permitted to carry out their managerial assignments without direct or indirect sabotage of their efforts.

part III

Theoretical Development

7 Some Theoretical Formulations Concerning the Processes of Social Change

Introduction and General Summary

We can now summarize and bring together into a coherent whole an exposition of the entire model and present some hypotheses about sociotechnological change that were generated by the application of the model to the analysis of examples of massive change affecting major institutions of large-scale societies. The substantive data we analyzed in order to illuminate various aspects of the entire model were derived from three studies. One concerned a complex modern technological system at an airport; the second, the largely successful attempt of nineteenth century Japanese society to incorporate European industrial and military technology; and the third, the total failure of traditional Chinese society to emulate the Japanese success.

Each study was undertaken to investigate and test some aspect of the model. The model as a whole was developed to meet the challenge of perfecting a holistic and integrated process model whereby technology and social organization could be treated within a single frame of reference. It was designed to accomplish the translation of substantive behavioral phenomena into a series of abstract, but measurable, variables of universal character. Such variables permit the comparative treatment in a single homogeneous field of the phenomenological contents currently treated as distinctive subject matters by different disciplines such as political science, anthropology, sociology, history, economics, etc. These variables, it is hoped, are an expression of the primary requisite that a true process model would treat organizational *formats* of empirical conditions rather than the phenomenological elements that comprise their *contents.* This shift from the substantive to the relational mode permits a universal and abstract frame of reference for the analysis of sociotechnological empirical conditions of enormous diversity and levels of complexity. Therefore, the empirical events and conditions of the real world, upon translation into variables that measure organizational formats, produce in this converted form a synoptic, analytical picture of sociotechnological interactions with great compression and economy of expression.

The model was designed, further, to deduce from the organizational formats derived from analyzing existing contemporary technological or sociopolitical conditions an extended series of format-arrays in the form of scenarios. Such scenarios represent the possible permutations and combinations of organizational formats that any type of sociotechnological change could induce as alternative states of organization for a given sociopolitical entity.

Underlying this model is a set of basic assumptions about the relationship of technology to sociocultural institutions and belief systems. Fundamentally, we treat technology and social institutions as aspects of a single cultural process and both reflect the same organizational requirements, viz., to arrange men and artifacts into patterned relationships to accomplish goals. A culture, in this model, can be viewed, ultimately, as comprising a series of tasks or activities requiring organization of men and materials to accomplish some end. Technology, for us, is the direct manipulation of the physical world by men and artifacts. Social organization is more concerned with the patterned relationships among men engaged in carrying out tasks in the real world. Actually, the management of technology is seen as the first level of social organization in the model. Where the tasks, due to their heterogeneity or complexity, require a series of managerial functions, a second order of social organization arises, namely, the need to "manage the management" of such complex organizational problems. Emerging out of this situation are organizational activities most characteristic of the political order (here defined as *policy* generation and maintenance) where numerous, very differently organized tasks have to be integrated or balanced to maintain the viability of a complex and heterogeneous pattern of society.

In sum, the model treats three zones of organization that interact to produce the sociotechnological configuration (*i.e.,* a "culture"): the Technological Zone, the Managerial Zone, and the Political Zone. Each of these zones must be analyzed for its organizational format in order that their combined interactions might be made explicit for analytic purposes. These formats emerge from the measurements of the variables that describe the behaviors characteristic of each zone of activity. The three zones are comparable in organizational terms since the variables we developed measure the same aspect of activity, namely, the degree of constraint operating on the interacting components that constitute each zone. The organizational structure of each zone is manifested in the interrelationship of three components, which it is the purpose of the variables to measure and describe. The three components emerged from an empirical and logical inquiry of a cross-cultural nature which resulted in the conclusion that to characterize any sociotechnological activity, there, minimally and universally, would have to be (a) entities that interact (men and/or artifacts), (b) a task to be accomplished, and (c) a setting which delimits the scope and boundaries of an enterprise.

Though the model can be used for specific analytic purposes involved in a single zone such as technology (chapter 4), the major concern in this chapter is to articulate the interzonal organizational properties of broader sociocultural processes under conditions of change stemming from any or all three zones in various combinations (since it is clear that sometimes technological changes are the "efficient" cause of change, and in others that ideological or cultural events are the sources of shifts in organizational formats). Analytically, we define change as a shift in organizational format.

The macro-level of analysis in the model is represented by the political sphere. Here we treat policy decisions that attempt to create organizational formats capable of making all three zones conform to a coherent pattern. However the model also can treat situations that express random or erratic modes of reacting to uncontrolled changes in the Technological or Managerial Zones. The political function, thus, is essentially the conscious (planned or unplanned) selection of alternative organizational formats that link the three zones. The model describes all possible organizational formats for zones and interzonal articulations from the "chaotic" unorganized to highly systematic and consistent organization modes. Such interzonal arrangements are termed *strategies.* Any behaviors describable in the model thus exhibit certain strategic properties which relate to their abilities to meet task requirements. In this form the model becomes an instrument for assessing or evaluating the success or failure *potentials* of the way in which a technological activity or social institution is organizationally constituted. Finally, once the total model's resources have been used to collect and analyze the conditions obtained in all three zones at a given moment in time, it is possible to produce an array of scenarios illustrating the various strategies of combinations, both inter- and intrazonally, available. The model makes no necessary assumptions of a normative nature. Scenarios for a strategy of reducing zonal or interzonal complexity are as readily displayed as potential organizational formats as those which project higher levels of complexity.

This chapter specifically seeks to make explicit the rules that ensure zonal and interzonal viability of the macro-system, *i.e.,* society. A viable system can take many forms. The application of the model in our pilot studies appears to indicate already that the organizational formats of any zone need not be entirely internally consistent, as in a classic organic model of tight functional unity. Many activities, even of a technological nature, in the real world require minimal constraint to insure adequate performance (*e.g.,* an economic activity like harvesting wild roots with a digging stick as contrasted with the organization of a steel mill).

As we shall show below, various organizational formats mixing both high and low constraint elements in the components that make up a zone is a typical feature of the real world. We eschew the engineering criteria of optimization or efficiency as an absolute normative goal in selecting a strategy. More concretely, the model allows planners concerned with potentially undesirable effects of technological change to select a strategic policy which opts for a lesser level of technological complexity in order to preserve more traditional, highly prized sociocultural features.

While this indicates the possible practical utility of applying the model, such studies should also produce a better basic scientific understanding of the processes of social change. It is hoped the model may shed new light on such classic problems as those which Weber made famous, namely, the origins of the capitalist social order in the Western world, how new social orders are legitimized, and the function or possibility of rational planning in human affairs. In addition, the rich anthropological literature that exposes the wide variety of sociotechnological conditions manifested by different tribal and non-Western complex societies in which they display different patterns of persistence and change of their institutions can be subjected to a new form of analysis different from the theoretical orientations now prevalent (for example, cultural ecology of Steward; the techno-economic evolutionism of White, Service, or Harris; or the Parsons and Bales psycho-social functionalist mode of interpreting change and continuity).

Characteristics of Zonal Configurations

What some writers call society and others culture is treated in this model as an interrelationship among the three zones. While there are phenomena of organization that can be analyzed intrazonally, macro-social phenomena must be modelled by the interface of the three zones. Thus a distinction between *intra-* and *inter*zonal analytic levels emerges. The use of the model requires recognition that though there are functional analogues in the organizational features of the three zones, their differences as distinctive analytic entities must be respected. Therefore, when macro-social issues require a synthetic analysis of their interaction, a new set of operating principles that model their interaction must be developed. The principles that describe component interactions within a zone cannot be applied directly to the more complex circumstances of interzonal interaction.

Principles of Intrazonal Interactions

Our first pilot study explored the organizational format of a single zone, the Technological, utilizing the highly complex radar approach system of an airport as an example. The scenarios produced in the analysis show the logically possible array of alternatives of organization for a technical system acknowledged to be in need of considerable change. From the total array of seven alternative scenarios of change from the existing system, only some of them appear applicable as possible effective solutions to the problem of uncontrolled aircraft entrance into airport airspace. From this study we learned the significance of the *degree of consistency* exhibited by the components that make up a zone. By consistency we mean the range of variation that may exist between components of a zone in terms of their measured constraint characteristics (low or high). A scenario represents one of the possible arrays of organizational formats of the three components of a zone. The requirement of a minimal degree of consistency among components results from the need for an organizational format to be capable of meeting a task requirement. The degree of consistency need not be absolute but the components must interact with *some* degree of equivalency of organizational format because there is a point beyond which different constraint patterns produce dysfunction.

The different normative judgments implied in the terms success or failure can be operationalized in the model as the degree to which a particular scenario of choosing organizational features can meet the requirements of some perceived task.

The consistency feature of a scenario is used to evaluate the potential effectiveness or ineffectiveness (*i.e.,* the success or failure) of an organizational strategy. In addition, consistency functions as a diagnostic instrument for assessing the systemic characteristics of a single zone, especially where dysfunctions among components are suspected but not clearly identifiable. The air traffic control study illustrates this use of the model as a diagnostic tool in which the Setting Component was found to be inconsistent with the Task requisites of controlling incoming aircraft without overloading the human and mechanical capabilities of the system. In other words, a potentially excellent system was being sabotaged by an uncontrolled component. One of the array of scenarios presented clearly indicates the corrective action which would produce more consistency and make the radar approach facility more viable. This study also exemplifies an obvious fact of sociocultural reality, namely, that many partially consistent systems *can* work, but that the price of inconsistency may be high. In this case, the human air traffic controllers must suffer highly stressful working conditions to make up for the ineffective organizational feature of the zone.

Principles of Interzonal Interactions

The second aspect of the model concerns zonal interface characteristics. As already stated, interzonal relationships do not require the same consistency characteristics that occur *within* any given zone. The Technological Zone, being tied directly to the physical universe, exhibits higher degrees of constraint required for successful activities than is the case in the realm of social organization, particularly in the political sphere.

The acts that move men are more arbitrary and variable in a normative world of ideas that characterizes the cultural milieu, whereas technology is structured by the nature of the physical universe itself which is reflected in the nature of the tools and equipment used by men in techno-economic productive acts.

But because interzonal transactions characterize the macro-order of society, a different class of interrelationships pertain, and we refer to these relationships by the term *correspondence.* Correspondence is determined by assessing the range of variation between two or all three zones with respect to their organizational format. Just as in the case of assessing intrazonal consistency features, a macro-organizational structure has viability if it can meet its task requisites within the tolerances defined culturally by the policy-making entity. Viability, interzonally, must be characterized by a certain required degree of correspondence in organizational terms. Despite the fact that viability includes an ideational normative aspect in which cultural attitudes and values figure very greatly, these cannot be absolute. Every culture must meet certain minimal standards of format articulation because many techno-economic tasks that make up the intrastructure of society *do* have rigorous format requirements of organization. When the ideational comes into direct conflict with the technological to the extent that men and resources are not allocated to such tasks, the culture disintegrates. Our Chinese study is a most clear exemplification of this principle. The nineteenth century Manchu Dynasty's desperate attempts to maintain the traditional bureaucratic structure of the empire and preserve the social order headed by the literati in the face of an acknowledged need to introduce revolutionary foreign industrial-technological complexes led to an insoluble organization impasse. Neither the Political nor the Managerial Zones, as traditionally organized, could effectively incorporate and maintain such new technologies, and the insistence of the central government that its own traditional agents manage the new enterprises led to disaster.

In sum, we have introduced the terms consistency and correspondence to describe conditions of similarity or dissimilarity of constraint characteristics within and between zones. Consistency is a measure of the degree of similarity or dissimilarity in constraint levels between *components* of a zone. Correspondence is a measure of the relative degree of similarity or dissimilarity in the total constraint characteristics *between* zones. Viability is a determination of the relative adequacy of an organizational format in relationship to a particular task.

The distinction between consistency and correspondence in the model is necessary because it is indeed possible for each zone of the model to display internally consistent values, which, nevertheless, are above or below those of one or both other zones. However, the concept of viability suggests that such organizational conditions will be associated with poor capacity to achieve desired ends.

Having defined our basic process terms, we can now concern ourselves with the systematics of their relationships as they define intrazonal and interzonal transactions. In terms of the model, culture or society as a system is an organization of ideas and social institutions which order or structure inputs into an array of organized activities in terms of men, artifacts, resources, such that the formats of their organization attain a requisite (not absolute) degree of consistency and correspondence with each other to produce a viable organizational order.

Processes of Culture Change

All cultures or societies, insofar as they persist as recognizable entities, must have the organizational capacity to absorb the impacts of new social or technological innovations whether independently invented or resulting from culture contacts. This section examines the variety of response formats open to any society undergoing change.

In the foregoing section we presented the basic terms required for the use of the model to analyze the process of sociotechnological change. We now proceed to the deductive implications of the model for the creation of hypotheses about the processes of change. Then, these hypotheses will be tested against the empirical data available from our three pilot studies. The overall procedure in the development of the total model thus included an inductive phase in which empirical techno-institutional case histories were utilized to develop the variables to be used and the procedures to be followed in translating data into abstract organizational formats, for only in this condition could the model be used for analytic purposes. In this form the model permits the generation of deductive hypotheses concerning the effects of different combinations and impacts of various organizational states on any given existing organizational structure (the model's terms for culture or society). These hypotheses are then subjected to empirical testing utilizing examples of the coded data derived from the pilot studies.

Obviously, in these research efforts we have only dealt with a limited range of sociotechnological phenomena, specifically, with complex Asian societies of the nineteenth and twentieth centuries. If the model has validity, its analytical properties must permit it to treat organizational states for the whole range, simple to complex, of sociocultural types. We deliberately selected relatively complex societies for initial testing on the assumption that a model that could successfully treat complex situations of techno-social organization can surely effectively encompass "simpler" examples of organization. In part, we also reacted to the general criticisms made by other social scientists, who treat complex Western societies, that anthropological data and theoretical explanations are limited to primitive or pre-industrial societies in their scope and are not helpful in examining modern society and its problems.

Within the limitations imposed by the research data used to develop the model, three fundamental issues in processes of change emerged, concerning which analytic and explanatory hypotheses are offered.

The analysis of change must address itself to: (1) the problem of how to explain what circumstances will cause changes of organizational formats to occur; (2) what array of possible responses by a social entity organized in a given format is available to it under conditions of change; and (3) what is the order of significance to rearrangements (change) of elements within the three zones of the model such that some components may be more critical for achieving a condition of viability than others.

CAUSES OF CHANGE IN ORGANIZATIONAL FORMAT

The model, being general in scope, can generate hypotheses about change in organizational formats where the components of the zones have high, low, or mixed loadings as measured by the variables. Change, in the model, is measured by shifts in the pattern of component loadings. Since the data base used in the construction of the model is restricted to societies where conditions of change involve technological and managerial characteristics of extremely complex nature, they represent variables carrying high loadings. Therefore, discussion will be limited to a type of situation where impacts on existing organizational structures represent cases of reactions to inputs that have high loadings (*e.g.*, adoption of European industrial technology into Japan).

In this presentation we do not treat that range of situations where complex organizational formats react to inputs (technological or cultural) characterized by lower loadings, *e.g.*, where technologically sophisticated industrial complexes can receive inputs from craft-oriented modes of production in certain aspects of their operations (cf. the Swedish case where a human team replaced a machine-ordered assembly line process to reduce worker-job dissatisfactions).

Within these limits, causes of change will refer to any technological or managerial innovation displaying high loadings (a high constraint pattern) that requires for its adoption a change in the organizational format on the part of the recipient entity.

The procedure to be employed is to utilize the eight organizational states generated by the model (and used in all the pilot studies to analyze the data) in order to consider the effects of innovations involving high constraint on these eight formats.

ARRAY OF POSSIBLE RESPONSES TO ORGANIZATIONAL CHANGE

The effect on any existing organizational state (*i.e.,* one of the eight types generated by the model) that an innovation of high constraint can have exhibit any one of three response patterns: (1) *steady state,* characterized by format continuity at the precontact level; (2) *amplification* of the organizational format whereby a single component's values are changed but the average for the zone remains, nevertheless, constant; and (3) *transformation,* in which the organizational format is fundamentally changed.

Examples drawn from the pilot studies will serve to illustrate these three response patterns.

Steady State: Our research data is weakest in illustrating this reaction to change. Nevertheless, the Japanese case study, while not encompassing an entire zone, did furnish materials which offer data concerning particular components in the Managerial and especially in the Political Zone. An example of the latter is found in the Communications Component of the Political Zone. The codings for the Tokogawa Shogunate show high loadings for all three variables. The same pattern of high loadings for all three variables is found in the Post-Meiji Communications Component (see Table VIII).

Amplification: The airport tracon room study furnishes a clear example of the need for amplification to improve an existing system. A scenario is suggested with this particular solution in mind. The diagnostic situation indicates a need to amplify (raise the constraint level) the Setting Component of the Technological Zone. The first scenario is one that changes only one component to create a more viable system format. In effect, a strategy to discipline and control inputs (*i.e.,* aircraft arrivals into airspace), a "Setting" condition, would permit the other two components to operate at full design capacity.

An example drawn from the Japanese case study illustrates this for the Managerial Zone, affecting the Legitimacy Component (Table VII).

Here the change of managerial capacity manifested by the feudal Daimyo when they assumed new managerial roles as industrialists (the Zaibatsu) shows movement in the Legitimacy Component, where the Locus of Validation variable has its loading changed from low to high (Table VII). Simply put, managing steel mills and steamship lines required an impersonal authority rather than the more personal, face-to-face basis for authority characteristic of the leadership in a small feudal fief.

Transformation: This involves a shift in total zonal format unlike the above two types wherein the zonal average remained constant despite changes in individual components. There are two subpatterns displayed by a transformational mode. One is where two or more of the variables change their loading so that the average value (low or high) of the components of the *entire* zone is changed. The other is where only *one* (of the three) component's value is changed, but the *average value* of the component or zone, nevertheless, is changed. Though our research does not yet substantiate a hypothesis that the greater the number of components requiring change, the greater the difficulty of inaugurating change, such a hypothesis is attractive on purely deductive grounds.

The first transformational subtype can be illustrated in the following cases. In Japan, the transformation of the managerial role of the Post-Meiji central government contrasted with that of the Tokogawa Shogunate is illustrated in all three components of the Managerial Zone (Table VII). Briefly, the Shogunate as a political entity had limited managerial responsibilities for techno-economic activities, whereas the new central government actively engaged in establishing industrial and military enterprises involving the management of technological activities. The Chinese example illustrates transformation in the other direction, especially in the Political Zone. In Table XII the imperial government displays uniformly high loadings for the Communications Component in 1860 and uniformly low loadings in 1910. This reflects the collapse of the Manchu Dynasty's control over the empire. What appears empirically as a form of disorganization is within the model displayed as a shift from one of the eight formats into another, indicating the ability of the model to treat conditions considered to be chaotic or retrograde.

The second transformational subtype is represented by the change of a single element which changes the *average* value of the component or zone. The Japanese case provides an example of this in the Political Zone.

Here we contrast the political roles of the feudal Daimyo with that of the industrialist Zaibatsu (Table VIII) which shows, for the Domain and Legitimacy Components, this type of transformation. The Daimyo, as the feudal head of a small agricultural and handicraft-oriented techno-social order, obviously would display low codings in the Domain Component. The greater complexity and scale of industrial enterprise managed by the Zaibatsu industrialists, through their intimate connections with the central government, shifted the value loadings to the high range.

We may now summarize in a table (Table XIII) the effects of the impact of highly constrained innovations on all eight organizational formats. In the table, the Initial State column represents the loadings of the eight organizational formats for all three zones. The next column, the Response State, indicates which of the three response types, steady state, amplification, or transformation, will result from the impact of an innovation of high constraint characteristics. By reading across the table the response pattern of each organizational state is predicted.

RELATIVE SIGNIFICANCE OF COMPONENT CHANGE FOR ZONAL VIABILITY

In utilizing three components to describe a zone, the model raises the issue as to whether all three components are of equal importance in their contribution to the viability of a given organizational state (an issue studiously avoided in classic structural-functional analysis). If so, a change in any component would be equally significant for changing organizational formats. If, on the other hand, some of the components exhibit a greater impact in their contribution to the structuring of the organizational whole than others, then the processes of change discussed above in terms of predictive capabilities of the model will lead us to assign certain analytic priorities to such components.

While definitive evidence is not available with reference to our data, there are some indications to suggest the hypothesis that in the Technological Zone the Man-Artifact Component is more determinative of the viability of zonal organization than the Task or Setting Components. This hypothesis emerged out of the current status of the airport radar approach system. Here we see that inadequate controls over the Setting Component inputs (random incoming aircraft) creates difficulties or ambiguities in the Task Component in that the smooth handling of computerized scheduled aircraft arrivals is disrupted by such random inputs. The system as a whole can still preserve viability (*i.e.,* handle this mix of task problems) by making adjustments possible only at the Man-Artifact Component level, namely, falling back on the ability of human

Table XIII. Zonal Consistency Under Conditions of High Constraint Innovations

	Initial State			*Response State*			
Organizational Format	*Man-Artifact or Communications*	*Task or Domain*	*Setting or Legitimacy*	*Man-Artifact or Communications*	*Task or Domain*	*Setting or Legitimacy*	*Tactical Format*
1	H	H	H	S	S	S	S
2	H	H	L	S	S	A	A
3	H	L	L	S	A	A	T
4	H	L	H	S	A	S	A
5	L	L	L	S	S	S	T
6	L	L	H	A	A	S	T
7	L	H	H	A	S	S	A
8	L	L	H	A	S	A	T

S = Steady state; A = Amplification; T = Transformation

controllers to override the automated aspects of the system (or, in the language of the model, accommodating by lowering the constraints operating on this component to meet the randomness [low constraints] of the setting). We indicated in presenting various scenarios for the airport study why it is doubtful that changing the values of the other components would yield a viable system capable of meeting the air traffic control needs.

The Chinese case yields evidence that for the realm of social organization, the most significant component, at least in the Political Zone, is the Legitimacy Component. This finding is entirely in line with Weber's classic analysis of the role of ideologies and social control as being central, rather than finding as the Marxian analysts have that the techno-economic aspects of culture are the dominant or leading aspects of a social order. The descriptive data in the Chinese study shows that despite the steady deterioration of the technological and managerial capacities of the imperial government, the regime was able to maintain itself through an ever increasing dependence on ideological grounds such as the maintenance of a valued moral order. Under conditions where they lacked effective coercive instruments of social control such as an army, militia, or police force, the empress' moral authority over the provincial gentry, the literati, and many village headmen was sufficient to preserve for decades the traditional bureaucratic organizational format in the face of a techno-managerial collapse.

Zonal Viability Under Conditions of Change

Since we have defined consistency in zonal organization as a viable state of organization, only some of the response types in Table XIII will meet this criteria of consistency. This will permit us to formulate empirically testable hypotheses about zonal viability under situations of change. A *steady state* as defined above is indicative of zonal consistency. *Amplification* responses are also "consistent" because the average value loadings of the zone do not change. The more challenging situation is where a *transformational* mode is required since it is here that inconsistency can occur. Since both consistency and inconsistency as concepts are operationally defined as ranges of variation measured by variables and components, they are inductively and empirically derived states rather than circular self-defined entities.

Therefore, of the eight possible reactions to change, four are potentially inconsistent and, as we shall show, only when the necessary accommodations are not made organizationally (*i.e.*, shifting component loadings in the correct direction indicated by the model). These can be located in Table XIII as lines 3, 5, 6, and 8. A diagnostic utilization of the model is demonstrated in the ability to locate among the components of a zone the *most probable source of inconsistency.* Thus, for example, in a case of technological change, which would fall into line 8, the most probable locus of inconsistency would be in the Man-Artifact Component since, as already pointed out, this is the most organizationally critical component of the two that require change in this transformational mode.

A social organizational example for the Political Zone, utilizing line 3, would most probably have its locus of inconsistency in the Legitimacy Component.

Even should the hypotheses about the priority of components in the three zones prove false, the concept of zonal consistency would still be valid. Since consistency is not an entity but a statistical statement of variations growing out of empirical measurements even if all components are of equal significance, it would still follow that components exhibiting too great a spread of their ranges of variation cannot interact in a viable manner.

Earlier, we defined a *strategy of change* as one aimed at achieving an overall condition of correspondence among the three zones of a macro-social entity. The lesser analogue to this situation is that of achieving zonal consistency as a *tactical* necessity since a successful transformational strategy requires the achievement of a certain level of consistency in each zone for total system viability. An example of a successful tactical, *i.e.*, consistent, operation in the Technological and Managerial Zones is found in the Japanese case study. As pointed out in our analysis, the Political Zone did not achieve measurements indicative of consistency; hence, at the strategic level, full correspondence among the three zones was not achieved. Table V of the Japanese case shows, on the other hand, that in the Technological Zone completely consistent transformations were achieved with all of the component's loadings moving to the high constraint features of a European industrial type. Table VI of the Japanese case indicates, however, a certain degree of *inconsistency* in the Managerial Zone. Since, empirically, management of the new industrial-military complex was shared by the central government and the Zaibatsu industrialists and bankers, we must examine the loadings for *both* groups to determine the degree of transformational success. While the Zaibatsu were completely successful in their organizational transformation, the central government exhibited inconsistency in each component, *i.e.*, a low loading for a single variable in each component where a high one in all three variables is required by the Western technology. This weakness is masked by the fact that the component average for the central government is sufficiently high to manage the new technologies *relatively* effectively.

In the Political Zone both the central government and the Zaibatsu exhibit weakness in the Domain Component. Briefly stated, this reflects the retention of feudal features of organization in which narrow, conservative cliques represented a repressive ruling class, an anomalous feature in a burgeoning industrial society where various new social entities are emerging and clamoring for power (organized labor, intellectuals and professionals, a commercial and industrial bourgeoisie).

Interzonal Viability Under Conditions of Change

All of the foregoing theoretical and operational procedures contribute to the final presentation of a process model capable of describing change in macro-social entities, *i.e.,* cultures or societies. The maximal powers of analysis of the model are seen when all three zones are modelled in interaction. In this section we attempt to close the gap between events in the real world, as described substantively by sociologists, political scientists, anthropologists, and historians, and the world of abstract theoretical constructs encompassed by such terms as culture, personality or society, and by such pseudo-process terms as "evolution", "adaptation", "pattern change", and "acculturation". Though we, indeed, utilize such terms as consistency, correspondence, and viability, these, as opposed to the others, are operationally defined in more rigorous ways as they are empirically grounded concepts that can be expressed by quantitative measurement (albeit crudely at this point).

As noted, a *strategy* is the term used to refer to the conscious or unconscious attempt by a policy-making (sociocultural) entity to respond to change which affects all three zones in the total model. Thus, the term refers to an analytical perception of the mode of achieving, or failure to achieve, *correspondence* between zones of organization, arrived through applying the model. Since the concept of correspondence is a statistical concept allowing for ranges of accommodative responses, it does not imply a mechanical, normative arrangement of equivalences in constraint characteristics of organization among zones. In other words, a simplistic, structure-functional model is *not* lurking behind this concept of organizational correspondence. Simply put, the Japanese could maintain a viable social system in which primitive agriculture could coexist alongside a complex industrial productive economic order, all enclosed by a neo-feudalistic, oligarchic political envelope. What are ideological nonsequitors (feudalism and capitalism) need not be organizationally incompatible per se.

The response patterns of steady state, amplification, and transformation used in the analysis of zonal consistency can also be utilized in treating the issue of interzonal correspondence. The analysis of interzonal correspondence is more complex than that pertaining to intrazonal organization. Since change can be initiated in any or all three zones, the specific zone which is the initial source of change is a matter of considerable import. As asserted earlier, it is much more likely that constraints operating in the technological realm will produce impacts on specific components of that zone that are more immediately observable (*e.g.*, cut the supply of raw material to a steel mill and it ceases to operate) than is the case with shifts occurring in the sociocultural realms where changes may be more diffuse and less immediately observed. (Bureaucracies are notorious for hiding near total collapse circumstances under an outward appearance of stability.)

Nevertheless, it is possible to measure effects of ideational contents upon managerial organization and, through this, on technological behaviors. The Chinese example shows how various traditional attitudes and values prevented managerial organization from being effectively applied to new industrial or scientific enterprise.

The Japanese case is one of a limited achievement of adequate correspondence among the three zones, whereas the Chinese case illustrates a dismal failure to achieve minimal correspondence among zones. At this point it becomes clear that matters of interzonal strategy fall into the area of policy making by political entities. The "management of management" is a question of treating the interface among the three zones.

In making a broad strategic analysis, the fundamental unit utilized in the model is now an entire zone. Previously, in dealing with a single zone, the unit of analysis was the single component measured by three variables. The value of the component comes directly from averaging the values assigned to these variables, which directly reflect empirical behaviors or conditions. Therefore, in a interzonal context, the units of analysis are now organizational constructs twice removed from the empirical world and, hence, operational procedures are much more deductive in character. We can, however, retrace our deductive operations through the two levels of abstraction back to the original empirical sources. In addition, all hypotheses deductively formulated are to be tested empirically.

Table XIV. Zonal Correspondence under the Impact of High Constraint Innovations

	Initial State			*Response State*			
Organizational Format	*Technology*	*Management*	*Politics*	*Technology*	*Management*	*Politics*	*Strategic Format*
1	H	H	H	S	S	S	S
2	H	H	L	S	S	A	A
3	H	L	L	S	A	A	T
4	H	L	H	S	A	S	A
5	L	L	L	A	A	A	T
6	L	L	H	A	A	S	T
7	L	H	H	A	S	S	A
8	L	H	L	A	S	A	T

S = Steady state; A = Amplification; T = Transformation

In the procedure for eliciting interzonal correspondence, we assign a single value loading to an entire zone by averaging the values of each of the three components for each zone. This enables us to treat the three zones analogously to the treatment given components in establishing the degree of zonal consistency. Thus, Table XIV shows eight initial typological system-states representing the organizational status of a society, as well as the total array of potential response patterns a society can make in reaction to innovations manifesting high constraint characteristics, as they are simultaneously operating on all three zones. As in the case of treating interzonal processes of change, we only have materials available to illustrate that limited situation of change wherein the problems confronting a society require accommodation to innovations characterized by high constraint.

Table X of the Chinese case shows that in 1860, the total zonal average for technology is *low,* since all three components display low loadings. The Managerial Zone, Table XI, also displays an average of *low* loadings for all components. However, in Table XII we note that the zonal average reflecting the political behaviors of the Manchu imperial government is *high.* These zonal loadings are located in line 6 of Table XIV. By reading across the line, we see that the model indicates that a transformational mode is the proper strategic response. This requires that one of the two zones which now code low must be shifted in the direction of a higher loading to achieve a *minimal* interzonal correspondence. In empirical terms, this means that under the impact of high constraint demands accompanying European industrial technology, the Manchu Dynasty had to transform either the organizational format of the Managerial *or* the Technological Zone as a minimal response. This would have permitted the Chinese Empire to survive as a viable polity while still not achieving that degree of viability in organizational format attained by the Japanese in the same period. The Chinese case illustrates the strategic failure to achieve minimal interzonal correspondence; as we saw, the Japanese case while displaying a tactical weakness, most glaring in the Political Zone, represents a strategic success in achieving interzonal correspondence nevertheless.

Unfortunately, the research data presented do not include examples of steady-state strategies equivalent to those produced for the analysis of single zones. Since the process of correspondence among zones in a transformational mode is logically the most radically complex of the modalities, we feel confident that examples of steady-state strategies will easily yield to the analytic powers of the model. Examination of Table XIV provides clues as to what types of situations will supply data for validating this hypothesis. Under the impact of high constraint innovations, only line 1 exhibits a strategy of steady-state response. The initial state of a total society manifesting a high average loading for all three zones is obviously characteristic of an advanced industrial society. In a speculative scenario, we would predict that further high constraint stemming from innovation will not produce a significant format change. Such a situation thus would represent a steady-state response. (A pressing necessity to adopt organizational formats of reduced constraint features in a modern industrial state would be an entirely different matter, perhaps requiring a transformational mode.)

In the above comparison of China and Japan, we pointed out that the Japanese did achieve a limited strategic success, which, upon analysis, is an example of amplification rather than transformation. We deduce this conclusion from the model by examining the average loadings for each zone (tables V, VI, and VII, Japanese case). Table V clearly indicates initial low loadings for all components of the Technological Zone under the Shogunate, and uniformly high loadings in the Post-Meiji era. The descriptive data for coding the variables of the Managerial Zone amply display the dominant position of the feudal Daimyo as managerial entities in the Tokogawa Era, while the Zaibatsu industrialists and bankers represent the dominant managerial group in the Post-Meiji era. Each group averages high for its managerial capacities and thus the zone codes high for both time periods. A similar condition obtains in the Political Zone with reference to the organizational capacities of the Shogunate as the chief political entity of the Tokogawa period, and for the role of the new central government in the Post-Meiji era. Again, both entities display high codings for this zone in the two contrasting time periods. The total societal organizational configuration for Japan in the Tokogawa period (the value of the three zones) was one of low, high, high, or line 7 of Table XIV.

The *response* column for line 7 indicates that the correct strategy to achieve correspondence is to *amplify* the Technological Zone. The loadings for the Post-Meiji period indicate that this is exactly what was accomplished by the strategy the Japanese reform government followed.

We believe that the empirical case study validates the ability of the model to assess the degree of success or failure of policies meant to change organizational structures either at the tactical (zonal) or strategic (interzonal) level. Tables XIII and XIV summarize the rules which determine feasible organizational states under the impact of innovations of high constraint. Simply stated, the apparently obvious conclusion that innovations which fall within the organizational parameters (as measured by variables) of any existing social entity should not cause it to change can now be demonstrated empirically. Without this new way of making organizational parameters operationally measurable, the degree of stress entailed in accepting innovations could not be determined. Thus, when the gap between the existing organizational format of a society and the technological or managerial innovation required of it is too great, only a transformational strategy is capable of maintaining a viable society. The ability to measure such conditions, previously verbally asserted as vague quantities somehow too much or too little, represents the true value of such a model.

We have probably carried this analysis of the properties of the model as far as our data permit. Nevertheless, certain exciting suggestions emerge from the foregoing analytic exercise which, though admittedly of a speculative nature, point to areas of future research. It would appear from our analysis of the Japanese case of effective adoption of European industrial and military technology that successful sociocultural adaptations are more easily made in a transitional situation where only an *amplification* strategy is required of the society. The basic managerial components required to accommodate to the European technologies already existed in Tokogawa Japan in the Daimyo-Samurai-banker components of society. The total zonal averages for the managerial and political realms were already high in the Tokogawa era. Even though the Technological Zone was characterized by a low zonal loading, the average of all three zones remains high. Only one zone, the Technological, required change and the power to effect this existed in the contemporary sociopolitical order. A totally opposite picture is manifested in the zonal codings for the Chinese situation. Here both the Technological and Managerial Zones coded low so that the average for the total societal condition was low. Hence, a *transformational* strategy was called for, but the power to implement this was lacking. It is therefore suggested that the capacity of a society to undergo radical transformation and remain viable can be quantitatively assessed with reference to two factors: (1) the constraint requirements inherent in the innovations and (2) the tactical and strategic responses that produce zonal averages that fall into the range of variation dictated by the model's rules for achieving positive zonal correspondence.

Evidence for the operation of these principles appear to be quite abundant in cases where sociotechnological changes were accomplished by violent revolutions accompanied by ideological strife: the wars of the Counter-Reformation in Europe, the English and French revolutions, and, more recently, the Russian and Chinese revolutions. Translated into the terms of the model, these are instances where the gap between the organizational formats of the traditional social orders and their supporting technological infrastructures, and those of the challenging, innovative new social and technical orders was so great that only a violent revolutionary restructuring of political institutions would produce conditions where power could be placed in the hands of managerial entities who could then pursue a transformational policy. Again, such a statement appears pathetically obvious, yet we assert that the true organizational principles underlying the obvious facts of history have not been demonstrated by a type of analysis which makes the empirical events conform to testable hypotheses about the nature of techno-social interactions. The rich data resources from historical records of revolutionary change can now be utilized in new forms which will permit the testing of our proposed process-model.

Hypotheses about Macro-Change Generated by the Model

A. All complex technological systems have managerial and political aspects that must be analyzed as part of a single field. Therefore:

1. The higher the measurement of levels of constraint in organizational formats available in technological, managerial, and political institutions, the more complex the tasks that can be undertaken successfully (*e.g.,* tribal societies cannot produce and maintain steel mills).
2. When levels of constraint in organization of all three zones are not equivalent, the general effectiveness of society to undertake any complex task is lowered.
3. The degree of incapacity of a society to achieve successful outcomes for tasks is in direct relation to the degree of variation in levels of constraint among the three zones, such that wide discrepancies imply predictable failure, whereas minor discrepancies imply relatively minor departure from optimal achieve ment.

4. Systems that empirically "work" where there are discrepancies in constraint levels between zones can only do so if:
   a. The society is parasitically drawing resources (men, materials, information, capital, etc.) from *external* sources (*e.g.*, an American Indian reservation, or a client-state or colony receiving foreign aid).
   b. The society has internal, redundant capital and other resources in some sector of its economy and utilizes its accumulated "savings" to carry inefficient or collapsing enterprises (*e.g.*, major "socialized" industries in Great Britain, or AMTRAK in the United States).
   c. Inefficient or ineffective programs of activities can be imposed on society for extended periods of time if ideological or police-state oppression can prevent victims from reacting to such harmful impositions.

B. Processes of change within complex techno-social systems are of three types: steady state, amplification, and transformation. Therefore:

1. Empirical innovations that are codable at the same level of organizational constraint are analytically equivalent (*e.g.*, a mule train, or a string of rafts are analytically equivalent, just as a plow and a chariot can be equivalent in "organizational" format).
2. Those technological changes that do not require new behaviors which raise or lower organizational constraint levels in the three zones will not have significant impacts on existing institutions. That is, considerable change can be absorbed by existing levels of organization: steady state (*e.g.*, extension of prefabrication techniques from ship building to industrial housing).
3. Technological changes which *do* force responses that raise constraint requirements in all the zones must do so to approximately the same degree, otherwise they will stress the capacity of society to incorporate them successfully. The *amount* of stress in the system is measured by the discrepancies found in constraint loadings of the zones.
4. If pressures to adopt innovations are such that all three zones must be radically affected to encompass the change, the likelihood of a successful transition is very small. The transformational strategy is the most difficult and costly form of accommodation and the one which is the most "politically" oriented (cf. Communist China).

5. Most cases of successful incorporation of technological change are accomplished by *amplificatory* processes. Many supposed examples of "revolutionary" change (*i.e.,* "transformation", in our model) will prove, via measurement in the model, to be amplificatory change when viewed in terms of organizational constraint patterns (*e.g.,* many Japanese preindustrial capitalists and aristocratic managers easily fitted themselves into the managerial infrastructure of the new industrial society).

8 The Model and Modern Systems Approaches

In previous sections we have addressed ourselves to accomplishing three major objectives. The first was to develop the capacity to translate diverse empirical data categories usually associated with various different social sciences into the model and to specify the necessary procedures (chapters 4, 5, 6). The second was to identify and define basic processes of change. These we have found to be three dynamic conditions: steady state, amplification/reduction, and transformation (chapter 7). The third was to develop the concept of viability to indicate what organizational responses best match a complex enterprise and how this is determined through the analyses of conditions of zonal *consistency* and interzonal *correspondence* (chapter 7).

By this point in the presentation of the model, we have given various demonstrations of the descriptive powers of the model with reference to real sociocultural conditions and thereby have only implied its dynamic features. This chapter seeks to make explicit in a more comprehensive theoretical statement the dynamics of processes of maintenance and change.

At the heart of our theoretical orientation is the concept of *viability* (just as *constraint* was the core concept to discipline empirical inputs). The particular concept of viability we utilize has validity in the very abstract context of the model. It is not meant to be applied directly to real sociocultural events. The real behaviors or conditions of societies, past and present, in various states of stasis or change appear in confusing guises that defy systemic analysis. Numerous examples of apparent violation of the principles of organizational viability presented here can easily be found. Obviously, societies such as Manchu China, Tzarist Russia, Bangladesh, or various American Indian reservations will exist for extended periods of time despite enormous organizational dysfunctions. Upon closer examination such cases must be treated as special cases, as displaying pseudo-viability, to be exposed in time as the failures they are. Failures or deficiencies in organizational capacity can be temporarily made up by ideological or psychological expedients, such as, for example demanding personal sacrifices from individuals or whole sectors of societies, or through applying brutal punitive measures to prevent immediate collapse of demoralized societies; or disintegrating societies may continue to survive though the expedient of receiving massive aid from other polities, directly by receiving food, clothing, and technical assistance, or indirectly by incorporating foreign enclaves of "experts".

Such ethnographic diversity of adaptive strategies tend to mask principles that apply to such empirical cases. These principles can only be discovered and understood by removing the level of analysis from the empirical data themselves to the abstract realm of a process model. This abstract realm permits the clear formulation of theoretical hypotheses about the dynamics of change applicable to all sociocultural phenomena by avoiding the pitfalls of explaining every special and unique case. No general model can survive the assault of such a crushing weight of sheer data which in specific cases are indeed so heavily affected by special ecological, historical, psychological, or ideological circumstances. However, the hypotheses generated by the model at this abstract level must be tested at the empirical level, ultimately. Hypotheses generated by analyzing unique special cases, on the other hand, will never approach the level of abstraction necessary for a cross-culturally valid model. We remind the reader that in chapter 2 we explained why hypotheses derived from analyzing any of the supposed component subsystems of culture, such as social institutions, personality, technology, ecological systems, ideologies, or linguistic structures, are all equally nonapplicable to the development of our sort of process model.

Our approach to utilizing the fundamental concept of viability is based on cybernetic concepts. The reasons for choosing a cybernetic modality are various. Since Norbert Weiner (1954) presented his first application of cybernetic concepts to sociocultural issues, numerous social scientists such as Bateson (1968), Parsons (1954), or Deutch (1952) have sought to develop their theories along these lines. The attractiveness of Weiner's approach lay in its promise of elucidating and explaining the hidden dynamic features under the complexities of social phenomena and in the possibilities of quantifying the qualitative data of the social sciences. Our review of their attempts to produce a general systems statement of social or psychological processes leads us to the criticism that they failed to bridge the gap between the sophisticated formal characteristics of the cybernetic formulas that provided the *tools* of analysis, and the empirical, substantive phenomena they used as data. Therefore, their results were as inconclusive and unconvincing in the end as results gained from using other more traditional forms of analysis (reviewed in chapter 2). We do not find the theoretical tools of cybernetic theory to be at fault nor do we refuse to undertake the exploring of real behavior with such tools. But we do assert that there must be established a midrange of conceptual contact between these two levels of discourse. Such a theoretical midrange is the purpose of the model. (We are reminded here of the classic statement of Merton [1949] calling for his fellow sociologists to develop such midrange theoretical constructs.)

In developing our model, we utilized four basic concepts from cybernetic theory. The first is the concept of *feedback loops.* These are relationships that serve to link the various elements that are interacting. The second concept holds that each of the feedback loops has a limited capacity to affect the range of behavior of the interacting elements. Third is that each given system condition (defined by the model, chapter 7), *i.e.,* steady state, amplification/reduction, and transformation, activates in its own distinctive manner different parts of the total feedback system. The fourth concept is that of *capacity.* Since the term capacity is a critical one in the model, its meaning must be made explicit. Capacity means a measurable ability to act. A feedback loop refers to a certain degree of capacity for component elements of a system to interact. Simple examples will serve: For the transportation of materials in a culture which only utilizes human carriers, whether the pattern of organization of the task involves single versus multiple units will determine capacity. In a similar fashion, the capacity of an electrical circuit to carry amperage is determined by the size of the wire and its conductivity. In both cases, continued increments of load can exhaust the capacity of the system to act.

Having defined the basic terms and concepts utilized in the model, we now proceed to link system conditions (that is, the three processes of steady state, amplification/reduction, and transformation) to their related cybernetic processes. Here we identify three cybernetic processes, each of which is activated by each of the three system conditions.

Cybernetic Processes

The first cybernetic process is that of *equilibrium.* This conceptualization of cybernetic processes is the most frequently utilized and, hence, best known in the social sciences. The work of Weiner and Bateson have been characterized by its use. In terms of our model, the equilibrium process means the following: The feedback loops and their capacity are equal to the task to be performed, while at the same time, the organizational format of their interaction is preserved intact. This situation corresponds to the steady state system condition.

The second form of cybernetic process is characterized by the situation where the feedback loops, linking the interacting elements into a given format, under change, preserve that format but the capacity of the feedback loops is increased or decreased. The processes involved are analogous to those outlined by Murayama (1963) in his discussion of the "second cybernetics". In terms of our model, these processes are characteristic of the amplification/reduction system condition.

The third cybernetic process elucidated here can only be labeled, crudely, by following Murayama's terminological practice, the "third cybernetics". In this situation, the capacity of the original format of feedback looping becomes overloaded so that a new looping structure is required if a viable system is to emerge. This obviously corresponds to the transformational system condition.

Figure 1 summarizes the analytic tools we developed to analyze the viability of sociocultural entities. We have been able to specify the particular cybernetic characteristics of various system conditions as to their *degree of viability* under conditions of change. What analytic possibilities are now available to us which are not covered by the application of our concepts of zonal consistency and interzonal correspondence? The

Figure 1. Cybernetic Characteristics of Various System Conditions

System Condition	*Cybernetic Characteristics*
Steady state	Equilibrium = loops and capacities are equal to the task while preserving the format
Amplification/reduction	Second Cybernetics = loops preserve the format; capacity fluctuates
Transformation	Third Cybernetics = capacity overload forces new looping format

material presented in chapter 7 located the *sources* of viability or nonviability as mirrored in the loadings of components of zones. The addition of the cybernetic modality now permits us to specify what operations are going on in and among these components of zones in a new dimension. These operations are seen in terms of capacity and changes in the format of the feedback loops. In our model this high level of abstraction (unlike such second-order abstractions as society, personality system, value system, etc.) can be operationally related back by stages to the lowest empirical level. This is possible by shifting back from the most abstract cybernetic level of analysis to the zonal component level and, finally, via the model's specialized variables, deal with actual empirical behavior or conditions.

At the zonal level of analysis only diagnosis of which component or zone was viable is possible; the cybernetic mode permits us to analyze the *processes* underlying any condition of viability.

Above, we suggested it is possible to reverse the process of increasing and refining abstract formulations, and retreat back to the original data inputs utilized in the model. A demonstration of this assertion is, of course, called for here. To do this we will reconsider the same data utilized in chapter 7 to illustrate the concepts of consistency and correspondence. As pointed out, our data illustrative of a steady state system condition was restricted to the variables of a single component of a zone. This restriction will not permit us to use such data here since our cybernetic modality requires a zonal level of analytic discourse to define the system condition. However, the concept of steady state has been utilized in social analysis by various writers so its meaning should be sufficiently clear. It is the classic system maintenance, balanced or structural-functional model, so ably presented by Bateson (1972) in his analysis of Bali as a steady-state society.

Since our primary objective in treating sociocultural phenomena is to understand the more far-reaching classes of change, the analysis of examples of amplification/reduction and transformation system conditions is more critical to an expositional discussion. Here the Japanese and Chinese data serve very well as they do provide data for a total zonal interactional analysis.

AMPLIFICATION SYSTEM CONDITION

An illustration of this system condition exists for the Managerial Zone in the Japanese case. Having established by our analytic discussion that such data fit an amplification system condition, we now attempt to *deduce* from the cybernetic characteristics of this system condition the means by which the empirical changes the historians reported came about. The cybernetic characteristics involve an increase in the capacity of the feedback loops that link the elements in interaction. More empirically, what does this mean?

We have described how a shift in the Managerial Zone occurred through the conversion of the Japanese feudal lords and merchant-bankers into a new industrialist (Zaibatsu) class. This shift was reflected in changes of loadings in the variables of the Legitimacy Component of the Managerial Zone. This locates the specific set of empirical circumstances wherein the change occurred. Our cybernetic model exposes that what happens in such a case is an increase in the capacity of the feedback loops linking the behaviors described by the Legitimacy variables to the Managerial Zone as a whole. What we need to know in any change analysis then is: (a) Is there a change in the capacity of the feedback loops, or (b), is there a change in the format of the linked loops themselves?

In this Japanese example, it is a clear case of only the capacity of the loops being changed. Empirically, this is exemplified by the shift from a face-to-face, personal set of relationships among the Daimyo and their retainers over to the impersonal bureaucratic mode of the Zaibatsu. Nevertheless, the modern managerial behaviors still take place in a social context of superior and inferior rigid class relationships. In both cases, the feedback format is one of *hierarchy,* albeit one is a personal, and the other an impersonal class of relationships.

Another vital analytic discovery about such descriptive material is that it illustrated a principle: that the cybernetic relationship (in this case, the increase of capacity) is true of the *entire zone* and not just of the Legitimacy Component. Therefore, there is no particular hierarchy of priority among components of a zone in effecting capacity shifts. While the initial source of change might occur as mirrored by a particular component, the effect, cybernetically considered, is extended to the entire zone. This may explain why it is that certain limited changes, which might only seem to affect a limited range of linkages, may nevertheless drastically affect the whole network. Empirically, the sources of change may be ideological or symbolic entities, in which case they may appear coded in the Legitimacy Component; or if they are technological changes, as in a shift from agriculture to industrial enterprise, their effects may be mirrored in shifts in the variables of the Man-Artifact Component. Still, such knowledge of the sources of empirical circumstances will not tell us how a society will respond to change, organizationally, nor can such information help us to decide which cybernetic process is going to be involved. It is only when we know what the model terms as the system condition that we can predict the effects of empirical inputs on the society's viability and adaptive capacity *as related to specific tasks.*

TRANSFORMATIONAL SYSTEM CONDITION

The Japanese data illustrates the system condition of transformation in the Political Zone. Historically, the political roles of the feudal Daimyo were limited in scope in comparison to those manifested by the Zaibatsu industrialist class. The cybernetic interpretation of the implications of such a change explains how this occurred. We have shown that the Domain and Legitimacy Components of the Political Zone underwent change. The cybernetic analysis shows that a transformational system condition occurred as a result of the overloading of the capacity of the feedback loops available to a society organized on a feudal, techno-managerial basis. Empirically, this reflects the historical conditions whereby the Zaibatsu became a nationwide entrepreneurial class with a high degree of responsibility, organized in an impersonal bureaucratic manner. Cybernetically, the capacity of the Daimyo was scaled to the small confines of a feudal fief. To manage the new national techno-economic infrastructure of a modern industrial state obviously overloaded the capacity of the original feudal format and this forced the creation of new loops, *i.e.,* new formats. The example of the Chinese attempt to preserve the old format under conditions of overloaded capacity illustrates the failure to attain viability utilizing their policy strategy.

Potential Utility of Other Systems Approaches in the Model

Our review of the manner in which social scientists (aided by the excellent survey of Buckley [1967]) attempted to relate the newer systems approaches to their standard subject matter suggests that their modus operandi was first to consider these sophisticated techniques or theories and then apply them directly to the subject matters of their concern. In other words, a marked deductive approach was involved because they did not first go through an intensive *inductive* phase of research activity that might have indicated selectively which of the many diverse systems approaches would best apply to the event-systems or conditions of relevance. This, we believe, was directly related to the underlying basic problem caused in the social science disciplines by viewing their data contents as related to substantively conceived entities. Therefore, the data themselves were not considered against the possibilities that they might not be usable in the frames of reference supplied by the new theories and models because only those data that have an organizational and relational character could be used as content in such models. Therefore, to apply these models to traditional classes of data produces a non sequitor. Our first task, therefore, was to produce a class of data categories truly suited to process-model requirements of expressing organizational conditions.

Our research efforts were governed by the need to create such a class of variables. This provided the necessary tool or technique of coding those events that other social scientists had described in substantive terms. With this tool we undertook to describe the dynamics of a complex man-machine system at an airport. We wished to see if the translation of descriptive data about the behaviors of men and machines into the new variables would permit us to design an explanatory model of the aircraft radar approach and landing system which would provide a coherent and logically consistent picture of its organizational structure. This empirical study, then, indicated that the new data categories provided by the variables did permit us to use a model in which *zones* of activities could substitute for traditional ways of describing linkages of men or machines as systems.

At this point, we did not yet choose any particular systems approach in the analysis or development of research, only the general theoretical concept of "organization" per se guided us. Having satisfied ourselves that the Technological Zone of activity (as a level of analysis) could be treated in a model of the sort we were developing, we then proceeded to determine whether or not sociocultural institutions could be included within such a model. This was absolutely necessary because of the fundamental theoretical assumption that technology and social institutions both interact to produce the phenomena we observe in studies of change.

Therefore, the next step was to design a group of research studies in which some new technologies would be introduced into a sociocultural system—in this case, the Chinese and Japanese experiences of attempting mastery of Western European technology. The results of these research exercises provided a set of data and organizational conditions, ranging from adequate to inadequate sociocultural responses to change, that required theoretical explanation. Still without resorting to specific systems models, our rough analysis indicated two processes were characteristic of how the model described real conditions of change. The first had to do with the manner in which the interacting parts (components) of each zone were organized. What was involved was a principle of consistency. Degrees of consistency, then, described adequate or inadequate organizational formats. Because all three zones are involved in an analysis of a total cultural response to change, the relationship seen as a process of accommodation among the three analytic zones must be analyzed in terms of the nature of the formats they each display and the degree of similarity or dissimilarity that exists among them. Just as in a single zonal analysis wherein degrees of consistency are examined, in a total interzonal analysis degrees of correspondence are examined.

At this stage of research formulation produced by an inductive program of research, we were provided with a model that explained changes in terms of principles of consistency and correspondence which are explicitly operationally defined in their derivation and are descriptively measured as to degrees of viability in given states of organization. From this we were able to *deduce,* finally, three basic modes of reaction to technological change: steady state, amplification, transformation. These clearly are cybernetic in character. Therefore, we now self-consciously examined cybernetic theory for the correct terminology. These three dynamic modes represent the operations of cybernetic mechanisms of feedback. The imagery of feedback loops and the inherent capacity to handle "loads" of such loops, which in the model are the formats, now serve very well to identify the sociocultural processes as cybernetic modalities.

On Applying Other Systems Approaches in the Model

Though up to this point, the model, as applied to the research problems of incorporating new technologies into societies, has only led us to consider and develop cybernetic systems approaches, we do not thereby suggest that only these are relevant to the study of sociocultural change. Obviously, no single model would be capable of using all the various systems approaches now available in the literature, such as game theory, information theory, decision theory, etc. In treating these other attractive and powerful new instruments of analysis, we no longer have adequate supportive evidence to offer to illustrate definitively how these would be used. We can only indicate programmatically how such approaches might be applied to the issues raised by the model and hope thereby to stimulate the necessary research required to develop their use, as embodied in specific research procedures and analytic exercises.

The model has three levels of analysis. In descending order of level of organizational abstraction they are: the interzonal and zonal level; the component level; and the variable level. We have already treated the zonal and interzonal level in this chapter. Some of the other systems theories and methods mentioned above seem eminently suited to the analysis of the two "lower" levels and their data contents.

At the component level, where the organizational meaning of the variables are summed in order to create a basic organizational analytic unit for describing a system-state, we believe such systems approaches as decision theory (Fishburn, 1964), relevance trees (Martino, 1972), and network theory (Mesarovic, 1964) are the most likely to be of use. Decision theory provides techniques for estimating the "costs" of making a particular choice among alternatives. Two examples, implied (though not developed) by the Chinese and airport case studies, will illustrate the potential applicability of decision theory to our model.

Assuming that decision theory could have been available to the rulers of Imperial China, how would they have approached the problems of incorporating modern technologies and yet maintaining the traditional regime in power? In the analysis of this situation in chapter 6, we indicated various alternative strategies that might have been implemented other than the ones actually acted upon. The application to these alternatives of a decision theory exercise might have clearly indicated to the policy makers the weighted costs involved in each alternative. For example, the decision to use foreign experts rather than rely on the limited expertise of the traditional bureaucratic civil servants could have been given a quantitative assessment of what sacrifices in goods, services, or social values would be entailed in making such a choice. Decision theory would, then, permit a tactical selection from among the array of alternatives by indicating which of these would optimize the chances of achieving a desired end—in this case, the mastery of Western European technologies. Unfortunately, in the actual case, no such procedure was available and they shifted from one alternative to another erratically and hence failed to follow any of them to a firm conclusion.

In the case of the airport tracon room involving the development of a complex semiautomated system to solve the problem of crowded airspace, a critical feature of system development concerns the relative weighting of roles assigned to men versus machines. At the moment, decision theory is indeed available to planners and is currently being used. It appears that the policy decision has been made to favor greater automation. Decision theory has been applied, unfortunately, only to a limited selection of alternatives concerning the machine (*i.e.*, computer-controlled devices) aspects of the total system, which, however, still include important human inputs as well. The advantage of our approach lies in its ability to supply a better balanced picture of the overall characteristics of the real system and thereby can prevent neglect of significant inputs. At the moment, decision theory if limited to an inadequate sector of total analysis of a system can only lead to an illusion of treating the problems affecting a system, albeit in a sophisticated manner. A sophisticated technique of analysis cannot substitute for an adequate formulation of problems, which can only be treated when an adequate picture of a total system is available. The decision to use more and more sophisticated computers, even if the selection among computers is arrived at by decision theory, will not eliminate the problems caused by overcrowded airspace if the computers only create even more crowding by their selective efficiency in handling scheduled flights and the consequent need for more human operators to complement or override the automated component of the system—an especially ironic condition since the automated system was designed to eliminate or reduce the human sector of inputs.

Two other possible systems techniques may be applicable in treating problems which in our model fall at the component level, namely relevance trees and network theory. A relevance tree is used by planners to determine the nature of the hierarchy of requisite inputs required to implement some policy decision already made—possibly through a previous decision theory exercise. For example, had such a relevance tree technique been available to the Japanese Meiji reformers, an imaginative reconstruction of how they might have proceeded to apply it to their problems of adopting Western military and industrial technologies would be the following: If the policy makers had decided to establish a steel industry, the first level of relevance analysis would concern the resource base, that is, the availability of iron ore, coal, transportation facilities, etc. The second level of relevance, then, concerns dealing with what resources of conditions would be necessary to supply the previously ascertained prerequisite resources. In this case, this would involve mining enterprises, training of engineers, geologists or communications operators, and so on. A third level is then reached which involves the consideration of an executive-management infrastructure needed to balance and coordinate all the above complex and linked enterprises.

Such a chain of prerequisites required by the chosen end of creating a steel industry is carried out through as many levels as the perception of the total task indicates. It forms, as it were, a latticework upon which the hierarchy of decisions in a policy formulation is structured.

An application of network theory to this product of a set of linked procedures given in a relevance tree is found in what is called the "critical path method." The function of a critical path is to determine the logical structuring of the tasks set forth by the relevance tree analysis. This is necessary because, logically, certain of the sectors of an array of delimited tasks must be accomplished before others are possible. The critical path provides the sequencing of these tasks in a program. The Japanese example provides potential materials for illustrating these two techniques. We will take the liberty of creating a speculative scenario of the process of industrialization, which, of course, cannot substitute for a serious documented study.

The Meiji reformers had decided to build an industrial society, the heart of which they believed to be a modern steel industry. We already pointed out above how a relevance tree analysis indicating a chain of prerequisite levels of mutually supporting resources is involved. Japan lacked the basic raw materials of iron ore and coal in the necessary quantities. These, however, were available in Manchuria and Korea. To attain these, a modern military establishment appeared to be the best way to insure the attainment of such resources, that is, through conquest. In other words, the policy makers decided that among the alternatives available to them to obtain the raw materials, military conquest was weighted as a superior alternative to buying iron ore or coal with monies or entering into a barter-exchange relationship with those countries that controlled the resources. Since Japanese society at the time did not have the requisite resources to equip and supply a modern army and navy, they did indeed initially purchase many arms from Europe. These were used to conquer lands on mainland Asia. As soon as the Japanese later succeeded in developing their own industrial capacity, they dropped their reliance on alliances with Western European countries which had temporarily gained them political security from their enemies and supplied them with those technological products they still lacked. The critical path aspects of such a policy formulation can be shown in the following logical sequence:

1. The selection of the military sector as the most important and immediate one to be developed, that is, to be modernized.
2. Utilize a temporary, protective phase of seeking political alliances with strong powers, especially England, to prevent other foreign powers, like Russia, from overwhelming them while Japan was still too weak to challenge the most industrialized nations but quite strong enough to attack the Chinese or Russians.
3. Following the successful conquest of Korea and parts of Manchuria, the policy makers felt free to invest scarce resources in more direct industrial development and expansion.
4. Once the industrial development was strongly established, the Japanese felt free to drop the defense alliance with Great Britain and then proceeded with an imperialist expansionist policy.

This critical path of decisions, while extremely effective in developing a powerful military and industrial base in a remarkably short time, had certain deadly consequences for the future which the planners might possibly have been aware of if their efforts had been informed by a more long-term assessment of the implications of what they had begun as a program of attaining initial, limited objectives. The very success of their military and industrial efforts led to greater demand for resources which necessitated ever greater imperialist wars until, in the end, they confronted the greatest military-industrial powers in the world.

This example indicates that the application of limited systems techniques to partial aspects of a total field of inquiry exposes planners to a failure to cope with unforeseen developments directly or indirectly related to their prior decisions. Development of models along the lines we have set forth, hopefully, can supply a means of insuring that a wider and more comprehensive overview of complex sociocultural situations can provide better resources for the initial development of long-range policies.

All the above modern systems approaches could not really have been utilized effectively in such exercises of analysis because the heterogeneity and complexity of all the various input entities and conditions involved—people, institutions, resources, ideologies, etc.—were too diverse in character to be linked in a model in which all of them could be treated in a single frame of reference. The tiresome, heuristic device of showing each of these input entities as black boxes linked by arrows to show interactive processes never could make concrete what was occurring during interaction. Our translation of all such substantive categories of inputs into relational and measurable variables is meant to provide such a unified frame of reference for process analysis.

The last group of system approaches to be considered as to their contribution to the effectiveness of the model concerns *simulation techniques.* Since our model interprets events in the real world as matches and mismatches between the variables of the three zones of analysis, simulation techniques will prove invaluable in producing an array of possible combinations of variables when their loadings change over time.

Though various simulation techniques such as KSIM, QSIM, or Systems Dynamics can also be used in modelling complex systems at the strategic level of analysis, we believe they fail to be entirely effective. For example, Forrester's (1971) classic study of a future, linked, world production and resource system examines the influence of various changes in rates of consumption of such resources as coal, oil, iron ore, etc., in relationship to factors of population growth, economic and industrial development, and standards of living. The variables utilized in such models are really only impressionistically selected without any explicit theoretical model to provide the criteria of selection. The variables in our model do have such explicit criteria for the selection of data given in the manner in which we formulate task characteristics and their relationship to the organizational potential of social entities that must carry them out. Once the task or tasks are specified, it should be possible to identify those resources, goods, or services that are linked to the well-specified tasks and thereby ignore all sort of hypothetical inputs which may only have ideological, accidental, or marginal involvements. A systems dynamics approach is most effective when the elements involved are correctly identified.

At least within the technological and social interface region specified by the model, the significant or essential units of interaction are clearly identified and a systems dynamics approach toward these specific entities can be very fruitful. It is a waste of time and resources to analyze and consider situations or phenomena that are irrelevant to the task at hand or are addressed to such undefinable or amorphous conditions that treating them as variables is a futile exercise.

The use of models of the sort we suggest may force planners to accept the need first to conceptualize the system or systems they are looking at in better theoretical terms before they choose variables or conditions to examine. There has to be some theory to inform them of what the variables are doing and why. The variables themselves have no meaning apart from the systems they express. Ignorance of these systems and their interactions cannot be overcome by adding more and more variables to any analytic stew. Systems approaches without good theories to inform them are valueless or even dangerously misleading if inappropriately applied.

Some Limitations of the Model

The limitations of this model, to the extent that our research applications permit us to perceive them, primarily stem from the design of any highly abstract, synoptic picture of dynamically changing complex systems. At this level of abstraction, the analyst is twice removed from the empirical phenomena of the world. The empirical "data" must be translated into abstract relational variables, and then grouped into various abstract patterns of organization. Therefore, as a planning instrument, the model cannot be used to predict what particular kinds of new developments of a technological, social, or political nature will appear in the future. It cannot interpret present-day conditions, in extrapolation exercises, to predict what specific problems will emerge from future technological or social developments. But it can suggest, by analyzing an array of options, which organizational strategies can most fruitfully be applied should such problems actually come into being. In this sense, it is an instrument for anticipatory planning.

What the model can supply, in advance of any emergent set of problems, is an indication of the probable *magnitude* of cost to society in adapting itself to specific solutions required to solve a problem. The magnitude of a problem is assessed by indicating the amount of change in existing institutional resources and capacities required to overcome it.

When a specific technological innovation is being contemplated, the model can be used to indicate whether the current technological, political, or managerial institutions are capable of absorbing that innovation. If not, the model can indicate the *degree* of change in these institutions required to accommodate to the innovation. However, the model can only specify the *parameter requirements* for planning organizational responses to change. It cannot specify what precise behaviors will achieve the necessary levels of constraint indicated as a required organizational response. It will depend on the practical knowledge and resources available to planners to determine what mix of entities and procedures "work" in the context of what the model specifies as a required set of relationships between technological and social factors. In other words, the target organizational conditions are specified, but the practical means for achieving them are not.

In one sense, this limitation of specification is not a weakness of the model as an instrument for planning change. Because the model does not differentiate between varied empirical behaviors that nevertheless produce the same configuration of constraint, it permits planners to choose from a range of appropriate alternatives the most practical alternative for achieving a requisite organizational condition. But in yet another context of application, this neutrality of the model toward specific alternatives, *does* limit its use for those aspects of planning where *value orientations* in society strongly favor certain alternatives over others. The model does not take into account these ideological components of real problems.

There are, in addition, certain *methodological* limitations related to the use of the model when only simple bipolar contrasts in constraint are used. Obviously, in treating those problems in which finer discriminations between levels of constraint must be achieved, the manual application of the model to data will become increasingly difficult as the number of units in the scaling of variables increases. This will require computerization of programs of analysis and increase the costs of application of the model.

Finally, data for the model required judgmental interpretations of empirical behaviors and conditions to select a value for the degree of constraint in a variable. Therefore, some analytic exercises will require an additional "Delphi-like" process to resolve problems of scaling, where technical or scientific criteria cannot be easily applied to observed or reported phenomena.

9 Theoretical Recapitulation

In view of the serious inadequacies of traditional social science theory concerning sociocultural change (chapter 2), the issue is not *should* we apply the new, and clearly superior, systems approaches in analysis, but *where* should we apply these theories. The problem of where to locate the processes being analyzed is critical if modern systems theory is to be used properly to explain how or why an organism or a social entity is organized in a given manner, otherwise, the demon of substantive approaches to understanding things in the world may not be entirely exorcized from even the newer analytic models. True, the entity being studied—a biological organism, or a social institution—is no longer conceptualized as a thing, but as a set of processes binding elements in a pattern of interaction. True, just as in the study of biological organisms, social phenomena are viewed as produced by systemic linkages between component parts (people, in this case) in which information transmission (via language or other symbol signals) triggers actions. The component parts, *i.e.*, people, are indeed viewed as complex, free, capable of choice, goal choosing, and goal inventing, etc. All of this is well and good. This does eschew the mechanical view of people as automata or socially programmed "pigeons" a la Skinner, or of social institutions as static, mechanical, or teleologically ordered entities which could never have produced the cultural phenomena we observe.

Further, it is also part of the newer approach to change, growth, and evolutionary theory in biology, psychology, sociology, and anthropology that the total field of analysis of living organisms or of social entities must include the category of "environment" because biological organisms or social entities are not closed systems reacting to other animals, plants, geological or climatic conditions in an adaptive mode in preprogrammed and unalterable ways (like Leibnitzian monads). The new ecological theory in biological sciences has profoundly affected anthropological theory in the field of cultural ecology. There is a growing anthropological literature demonstrating how the economic order is related to the social and political order and how these, in turn, are linked to transactions with the physical environment, as mediated through technologies of production, *e.g.*, agriculture, pastoral, hunting, or industrial modes.

Therefore, we cannot locate our study of sociocultural phenomena in either the social entities themselves or in the bio-geographic conditions (plants, animals, climatic, or geological aspects of an "environment") these interact with. We need not expend our energies in interpreting what is going on, located as processes, occurring *within* the social entity that is adapting to the objects, entities or conditions in its environmental field; nor need we analyze or interpret, similarly, what is the nature or characteristics of the entities, biological, geological, or climatic, involved in a field of interaction in order to link social entities to what they are reacting or adapting to. To do this is to fall into another subtle, well-disguised form of substantive approach, hidden as such by the new process language used to describe all of these entities or conditions. Even dynamic process analyses of such entities are still making them the central focus of conceptual ordering. The true locus of analysis, then, must be *displaced* from the entities and their interactions, both viewed in an internal context (within an organism, biological or social) or in an external context (their relationships to other entities). The focus must be placed firmly upon the pattern of organization that the entities which interact display. It is not necessary to know in such a model why or how a given social entity is organized in this or that manner, but only what its capacity to carry out certain organized acts are as a result of its given mode of organization. It is not necessary to know how or why that which the given social entity is acting upon or with (other humans, institutions, tools, animals, plants) is organized or what are its own internal organizational characteristics. Only the pattern of interchange between all the entities that are interacting is the focus of analysis for our model. Hence, a radical *externalization* of the location of analysis occurs—external to the entities themselves. All the entities involved are conceptualized as making *inputs,* each in its own way, to the resulting design. How or why they make such inputs is not part of our analysis: only the *form* (capacity as measured by constraint features) of the inputs is required.

This is accomplished in the model by making the central feature of analysis not the entities that interact, but the *tasks* (organized activities that produce concrete results) that the involved entities carry out, explicitly or implicitly, directly or indirectly, as actors or as constraining conditions. This is possible because the model does not make the potentially lethal conceptual separation between social entities, machines, or environmental conditions, that is, add or link them as separate entities in interaction. To do this is to revert to a form of substantive analysis in which these entities are classified as different.

The solution to this difficult problem of showing interaction among entities of different nature and qualities (people, institutions, material objects, geological or climatic conditions, etc.), and yet, not allowing their distinctive "natures" to create the impossible philosophical task of explaining how and by what mechanisms they contact and affect each other is to be found in the invention of our new class of variables. In the Man-Artifact Component of the Technological Zone (which refers to activities related to direct task accomplishment), we note in the first variable (Locus of Dominance) that it is not necessary to analyze, define, or describe either the men or artifacts but only the pattern of behavior that emerges from their interaction. Whatever the nature or potentialities of human beings, the degree of constraint that the use of a tool or machine has on the behavior of the human is the critical factor. All the rich psycho-biological potentiality for behavior of a human actor is irrelevant, analytically, in the analysis of an empirical situation. True, an actor potentially has the choice of refusing to act—but if he does act, the resulting behavior is conditioned by the pattern of the interplay of man to men and these to their tools and machines or weapons. Our model is not concerned, therefore, with theory that explains motivation (though such is a legitimate enterprise for social psychologists).

Another example is to be found in the Setting Component. Here we can treat *selectively* those features classically called "ecological" or "environmental" without analyzing in a substantive mode the input characteristics of climatic or geological conditions or of floral or faunal entities required for the model's operation. Whatever entities or conditions exist in the real world surrounding the carrying out of a given task, such as plowing a field, building a pyramid, or driving a herd of buffalo over a cliff, only those aspects of such entities that actually come into play or create organizational problems by their presence need to be considered. In examining the variables of the Setting Component, we notice that some of these have to do with determining whether the entities or conditioning elements in the "external field" come into the arena of interplay in formats that are relatively amorphous or random (you are hunting a capricious animal that might or might not be there) or, on the other hand, in formats that are highly organized (the enemy is advancing in tight formations at a fixed rate). Thus, ecological factors become translated into dynamic conditions that are measured by variables. There is no need for special variables to describe hordes of rabbits versus armies of men, or of monsoons versus temporate rainfall conditions per se. In some contexts, a rabbit's behavior contextually may be measured by the same coding value as a human's or even some physical condition like a type of rain. All these diverse behaviors may be "random" to the same degree, though emitted by different entities.

A similar solution has been found for dealing with the problem of what to do with social institutions. Substantively considered, patrilineal or matrilineal kin types are different; the headmen of a tribe in council represent a social entity different from a board meeting for industrial executives, just as the Roman Catholic Pope and his cardinals are distinguishable from the Manchu Emperor and his heads of ministries. In the Managerial and Political Zones of the model, the three variables of the Communications Component suggest that the format of man-man interactions, in itself, has certain definitive consequences for capacity to perform certain acts. It is the capacity to perform such acts in relationship to specific tasks that is what truly matters and not whether that capacity is associated with a matrilineal kin type in one case and with a patrilineal kin type in another.

It becomes possible in this context to recognize that the similarities and differences that classical anthropological categories of classification assert may be irrelevant. The ways in which hunting peoples carry out complex hunts may be identical in format with those used by advanced industrial societies in organizing a steel mill. Both involve cooperation, complex communications, hierarchy of leadership, division of labor, etc. Hence the classification of hunting versus industrial society as a broad rubric is misleading in organizational terms. Similarly, the even more crude distinction between primitive and civilized people blinds one to the possibility of highly complex components of behavior in primitive societies and very simple ones in civilized societies.

Again, in the Domain Component of the model we see now the variables used to rescue us from substantively considering what is classically termed the "resource base" of a society. Usually, a sociological or anthropological discussion of a resource base involves analyses of technologies, the nature of various resources used, the modes of production applied, etc. In the context of our model the Size-Quantity variable, for example, can be coded with inputs from any kind of technology or mode of production. In the context of carrying out a task like building a pyramid, the resource base must be far greater than that required to erect a family granary. Such factors as scale, scope, and complexity in different formats are *measured* as such by the codings given to variables, and not merely implied or given qualitative assessments of greater or lesser degrees. The amount of flexibility or capacity in any given entity to meet organizational requirements can be seen to be related to its *redundancy* capacity which is now measurable by the variables in this component of the Managerial Zone.

Finally, in the variables of the Legitimacy Component of the Political Zone, we can see how the problem of avoiding identifying and treating distinctively all of the institutions in a given society is accomplished. If all we need know about a social entity is how it marshalls its resources in men and materials to carry out specific tasks, then only certain areas of information about a culture are required. A complete list of a society's institutions need not be amassed. There is an inbuilt selection of those social units or entities required for analysis, since the tasks themselves empirically determine what persons or artifacts, attached to which social agencies, are at work, directly or indirectly. And, of these agencies or entities, we need only know in what *form* they are making their inputs. Only information needed to code the variables, in other words, is involved as content for an analysis. Neither the total culture, nor even every aspect of a single institution need be involved in the analysis of a real situation.

In terms of more general theoretical concerns, the model begins to suggest an interesting theoretical explanation of why the organismic approach to functional integration or the functional utility of all parts of a society is unacceptable. First of all, once a minimal capacity to carry out the array of tasks confronting a society is achieved, there is no reason why added capacity or redundancy of resources (mental or material) would cause any harm. Given the human propensity to elaborate ideas and behaviors, conditions of leisure, surplus time and energy can generate and support all sorts of nonproductive actions or programs, including organized activities in aesthetic, recreational, or religious contexts.

Also, we need not worry about the consequences of a lack of total integration of all parts of the culture because only that *degree* of integration that is required for meeting task requisites is critical. There is no theoretical reason to become concerned about all the inactive or nonactive components of a social system, for example, survivals of past behaviors, borrowings from other societies of material objects or behaviors for aesthetic or status reasons, etc. Many institutions or ideas can abound and yet need not be operative. Their existence can be tolerated *if* they do not actively interfere with those behaviors that *must* be organized in a certain way to guarantee completion or success of essential tasks.

Another explanation of the existence of apparently overelaborated institutions, like bureaucracies in government or the military, is that there are sometime or occasional tasks which require more resources and organization than others. A society must be prepared with organizational formats for such tasks even if they are not everyday occurrences. In other words, institutions may be designed for maximal effects even if these are only applicable on limited occasions. A good example of this is a modern national army. In peace time, or during wars, between offensive operations, most of the men and materials may be idle or unused, but the bureaucratic heads design an organization capable of launching massive attacks to be sustained over large periods of time. In complex civilizations we see this *latent* aspect of organizational capacity wherein goods and resources are being hoarded by governments, industrial enterprises, or military establishments as a necessary condition for certain classes of action. This tendency, of necessity, leads to the fascinating consequences or requiring extreme social control systems to guarantee the flow of goods and services in the right channels to achieve conditions of vast latent capacity. The need of complex states for surpluses of all sorts becomes clearer given the nature of the economic, political, and military enterprises they engage in, *e.g.*, building pyramids, fighting wars, maintaining irrigation complexes, etc. Such withholding of resources must be validated ideologically or insured by direct use of punative force. The variables of the Legitimacy Component were designed to offer tools of analysis to measure how different types of social control devices, ideological or physically punative, operate and to measure how much control is required in terms of the needs of constraint placed upon populations to enforce a political policy, for example, to get masses of people to work for a generation on erecting a cathedral or a pyramid, or agree to enter armies to fight for years.

Perhaps we might suggest that all the above considerations lead to a formulation about the nature and degree of integration of cultural "systems"—namely, that the type and degree of integration of social institutions found is a calculus involving different degrees of constraint required of men and resources in order to produce an effective format to permit a society to carry out that array of tasks that empirically make up the routines of behavior manifested by any specific society. Some tasks require more coordination and resources than others. But all in all, what we call *culture,* as a body of ideas, refers to a repertoire of tactics and strategies for allocating goods and resources in given patterns to accomplish such tasks.

At this point, the more humanistically oriented reader will become dismayed with what appears to be a blatant disregard for the most significant "function" of culture, that of specifying the goals of human activity and the nature of the shared values, norms, and ideals that discipline members of society in their behavior. To such criticism we can only reply that this particular model and the theories it generates cannot treat such issues because to do so would once more place us in the position of defining or analyzing events and processes that are *internal* to the social entities or actors themselves. Our model can only treat event-systems *external* to the behaviors of people, alone or in groups. It tells us nothing about the mechanisms of information transfer, or group formation and solidarity, or the genesis of technologies.

The model, however, may tell us something significant about the general problem of change and evolution. The concept of adaptation of a society to an environment can be treated in more operational terms once translated into a frame of reference whereby the social field is linked to an environmental field via an analytic focus upon the tasks that conceptually unite their interactions. It now becomes possible to talk about the viability of societies in their given environments—as measured by "task effectiveness". We can learn nothing about the *mechanisms* whereby societies invent bodies of knowledge, technologies, or social institutions in such a model. The study of adaptive strategies at the mechanism level will have to be treated by biologists, psychologists, and social theorists who treat organisms, personalities, and interpersonal fields in terms of the new cybernetic and communication theory models. We readily agree that to understand how and why evolution from nonhuman to human cultures, from simpler to more complex social systems occurred, we have to posit that the development of complex personality systems is necessary to produce and maintain new forms of social orders, and that increasing conventionalization of gestures into full-blown symbol systems are also prerequisites to such changes. But we do argue that such information is not necessary for the development of the type of model we presented and the limited aims of explanation we sought for linking technological and social sectors of a culture.

10 Epilogue

Some Implications for Social and Cultural Theory

It has not been the purpose of this work to produce a new general theory of society and culture. The complex nature of cultural phenomena cannot be explored in all its dimensions by a specialized tool of analysis which can only show what can or cannot happen within general parameters if a social entity is organized in one of a series of patterns. It cannot explain how the enormously diverse sample of world cultures originally got organized in this or that manner. Only an analysis that takes into account man's complex relationships with the world around him and the world within him can do this. The model cannot suggest explanations such as the archeologist, the student of culture contacts, the explorer of human psycho-biological characteristics, analysts of kinship or value systems, and various other specialists offer as to how the cultural world arose, diversified, and maintains itself. It can only translate their rich data into dry abstractions of a universal character which describe their organizational attributes. But at least it can begin to tell us what is or is not possible for entities to accomplish if they are organized in this or that manner. If such a model indeed does have the capacity to assist social planners in the development of alternative programs of change for societies that wish to choose a certain direction of change and will point out among an array of possibilities the most favorable strategies to pursue, here is something of value. To the basic issue of whether or not such a society has wise leaders, the physical resources, the good fortune, the social cohesiveness, and motivational capacities necessary to achieve their goals, the model cannot address itself. Like a map to a wanderer, it may point out a route but the man must do the walking.

Though the testing of the capabilities of the model has led us to use illustrative materials that represent practical problems of incorporation of complex technologies in Asian societies, working with the model has inevitably stimulated certain thoughts and speculations concerning many basic problems in interpreting sociocultural phenomena. One such issue which has generated much debate among social theorists concerns the existence and functions of formal institutions and ideal rules of conduct and the relationship of these to actual behavior. The so-called structuralists stress the importance of kinship, legal, political, or religious institutions as embodied in systematic rules of conduct and thereby assert the priority of the social system in analytic exercises (Durkheim, Radcliffe-Brown). Writers such as Barth or Leach reject this position and assert that analysis of social phenomena should focus more on potent situational determinants located in ecological, social or personal, motivational factors. Geertz, apparently, favors stressing symbolic or ideological factors in forming human responses to conditions of change. Indeed, while not denying the existence of kinship, legal, political, or economic institutions altogether, descriptions of these as formal systems binding on behavior have been challenged as social scientific mental constructs not really existing as such in the minds of members of a given culture. Even if they do exist as part of the mental content of members of a culture, it can be shown by much ethnographic evidence that rules can be ignored, reinterpreted, bypassed, or used on special occasions only.[1]

It is our contention that none of the above points of departure for theoretical explanations, while perhaps very valid from certain substantive or empirical standpoints, utilize the proper conceptual level of analysis. The insistence that the systemic properties of social institutions directly affect human behavior and the counter suggestion that the actors themselves, as motivated by various situational imperatives, create and manipulate, variably, social systems are both inadequate theoretical stances. There is a missing analytic level between real acting individuals and cultural rules which the model seeks to supply. By choosing as our conceptual point of departure managerial modes of organization as arrayed against the specification of tasks, real behavior is only significant in the model when it refers to, and is contextually defined by, concrete tasks. Cultural idea systems or social institutions are only significantly put into their proper context when these actually allocate or route men and resources as managerial inputs into a given task. Now in this form, the model can relate people's behavior to cultural institutions at the appropriate level of analysis.

Unfortunately, investigating this key relationship between people's behavior and cultural systems has been made a confusing enterprise by the vast amount of ethnographic information on the richness and diversity of these cultural forms and behaviors. Thereby, the humanistic aspects of the discipline of anthropology has been treated more fully and effectively than scientific explanatory ones. The cultural ecologists (Stewart, 1955; Vayda, 1969, etc.) have already pioneered the way to a better scientific, rather than humanistic, interpretation of the development of culture by concentrating on those tasks—the techno-economic ones—which do inherently expose linkages between the carrying out of organized activities and a society's organizational characteristics and capacities to do so. Considerable achievements have been registered in interpreting cultural processes involving simpler hunting and gathering societies. Unfortunately, in dealing with more complex societies their

[1]This variability of real behavior from norms brings up the fascinating problem of why, if this is indeed so, the institutions and rules affecting these areas of behavior have an arbitrary and systematic character. At least two kinds of answers to this have been offered: (a) Levi-Strauss's (1963) thesis that to be intelligible and communicable as a set of rules they have to be fully worked out and logical; and (b) that formal rules serve primarily boundary-maintaining functions: that is, the rules set limits to variability of behavior and provide guideposts to control extremes rather than direct all aspects of actual behavior.

methods seem to lose interpretive powers as all sorts of *non*-techno-economic factors seem to have potent causal attributes.

In the area of religious, social, or aesthetic behavior, and some aspects of political activity, the associated tasks are indeed more open-ended or amorphous in character as compared to techno-economic ones, and less sensitive to immediate failure if not properly conceived or executed. In these rich zones of behavior it can easily be pointed out that despite such behavioral open-endedness institutions may yet display highly organized rules concerning behavior and impose strict sanctions on their observance, just as is the case in such technological behaviors as building a pyramid or manufacturing steel. Also, it has been pointed out again and again (*e.g.*, Geertz, 1963) in criticism of some cultural ecologists (*e.g.*, Stewart) that their concern for the so-called primary inputs of techno-economic factors as determinants of sociocultural phenomena does not fit the ethnographic facts which demonstrate conclusively that ideologies, social patterns, and culture contact situations all are also dynamic inputs in given situations that are as determining in their impacts as the techno-economic ones. Such observations tend to complicate if not obscure certain underlying principles of organization exposed by the model which can be seen more clearly in circumstances where to be *under*organized for a given task leads to disaster. To be *over*organized, *i.e.*, to have institutional capacity beyond what is required for effective performance, or to engage in elaborately organized kinds of activities in religious or aesthetic areas where organizational capacity per se is not an issue are both characteristics very commonly found, ethnographically. Cultural entities tend to display considerable redundancy and be tolerant of such unused capacity or proliferate non-productive but organized group activities (art, play, religion, etc.). But, nevertheless, there are those critical areas of behavior where unless the organizational format does match the task requisites, they cannot be carried out. This would be especially marked in the technological zone.

Therefore to provide evidence for the validity of our theory of organizational constraint, we must look to the full array of tasks (here to be distinguished from an even more general category of organized activities) of a given culture and assert that in those areas of task-oriented behavior where controlled and disciplined behaviors *are* required, the pattern of organization of the task must be matched by the complementary pattern of organization of the managing entities. In the language of the model: the formats of the zones must correspond.

What happens when we find areas of a culture (and we inevitably will, especially in complex societies) where there are many forms of behavior and tasks that are open-ended and situationally variable in their organization and resulting outcomes, yet the institutionalized rules concerning them are highly formal and strict? In such cases we can suspect secondary considerations are at work, such as:

1. They serve useful boundary-defining functions and do not control all aspects of behavior.
2. They may be needed to control certain limited but significant aspects of the total situation.
3. They are utilized and applied only on special occasions, and then, selectively.
4. They are the result of culture contact and borrowing from another culture, perhaps for prestige purpose; or, these may be forced on a society powerless to reject them (*e.g.*, colonial situations).

In other words, all the above empirically verifiable situations will be considered as special cases of the general rule that controlled institutional formats reflect association with disciplined and controlled organizational features in carrying out specific tasks.

Another issue on which the model can help bring interpretive clarity concerns value judgments in comparing societies. It becomes clear why we cannot consider complexity or heterogeneity of a culture, when comparisons are made cross-culturally, as indicating (as the nineteenth century evolutionists held) organizational superiority—if we measure such superiority in terms of the adequacy of institutional formats to provide the managerial inputs that match the performance requirements of the culturally relevant tasks. (Different cultures obviously have different arrays of tasks.) Whether or not a highly complex society is capable of coping with its organizational problems is determined in the same way as for a simple hunting band or agricultural village. Complex societies like Tzarist Russia, Manchu Dynasty China, or modern Bangladesh may be as badly organized culturally to cope with their confronting array of tasks as American Indian tribes on reservations or a hunting band in the Amazon Basin on the verge of extinction. According to the model, when tasks and organizational (managerial) capacity are out of phase, for whatever reasons (ecological, ideological, political situation), regardless of the richness and complexity of specific cultural institutions and the loyalty or veneration they generate, there are only two options as alternatives to cultural collapse: (1) give up or reject the given task or tasks you cannot cope with; or (2) develop (or borrow) new organization formats.

Sometimes this basic choice is avoided by three rather common strategies:

1. A society can avoid making a bitter choice, for a time at least, by denial of the situation, bolstered by a willingness to suffer the harsh consequences of a refusal to change. This is a possible strategy for some societies that have enormous moral or psychological inner resources which permit them to make up at a displaced psychological level for deficiencies in their actual technological or social organization. They can continue as coherent societies, filling the gap between new demands for performance that their current organizational capacity cannot match by the willingness of people to endure suffering and make heroic special efforts or accept psychological rewards in place of more materialistic ones.
2. Temporarily, some of the redundant capacity in the total social order which has not been exhausted can be allocated from one area where it still exists over to another where the capacity has vanished (*e.g.*, squeeze the peasants and merchants yet further when the bureaucratic political order wastes its resources building palaces or waging futile wars).
3. It is possible to make special arrangements with an external system to supply the society with those components of organizational capacity which are absent. Reservation-based tribal societies or some developing countries fit into this category.

We believe that though such solutions or adaptive strategies as the ones described above may appear in some cases to be effective, they may also disrupt efforts to face the fundamental issue of establishing a viable and stable society. Ultimately, this condition can only be achieved by an effective correspondence between the technological and managerial behaviors. Revitalization religious movements, for example those described by Wallace (1956) among American Indians, may temporarily assuage the hurt and hunger of demoralized populations undergoing overrapid change involving the loss of a cherished way of life. They may have the desirable effect of permitting people to cope with the psychological problems that are exceedingly stressful, but these turnings toward ideological solutions cannot long be effective if they do not lead to a development of the managerial controls over the parts of the real world that actually supply the material necessities of life a population is dependent upon.

bibliography

Ackoff, Russell L. "General System Theory and Systems Research: Contrasting Conceptions of Systems Science." In *Views on General Systems Theory,* edited by M.D. Mesarovic. New York: John Wiley & Sons, 1964.

______. *Fundamentals of Operations Research.* Chicago: Aldine-Atherton, 1968.

Adorno, Theodore et al. *The Authoritarian Personality.* New York: Macmillan, 1950.

Barnett, H. G. *Innovation: The Basis of Culture Change.* New York: McGraw-Hill, 1953.

Barth, F. "On the Study of Social Change." *American Anthropologist* 69 (6): 661–669, 1967.

Bateson, Gregory. "Redundancy and Coding." In *Animal Communication,* edited by T. A. Sebecok. Bloomington: Indiana University Press, 1968.

______. *Steps Toward an Ecology of Mind.* New York: Ballantine, 1972.

Beckey, George A. "The Human Operator in Control Systems." In *Systems Psychology,* edited by Kenyon B. DeGreene. New York: McGraw-Hill, 1970.

Beckmann, George M. *The Modernization of China and Japan.* New York: Harper & Row, 1962.

Benedict, R. F. *Patterns of Culture.* (orig. 1934) New York: Mentor Books (New American Library), 1956.

Boas, Franz. *Race, Language and Culture.* New York: The Free Press, 1940.

Boulding, Kenneth E. "General Systems Theory—The Skeleton of Science." *Management Science* 2:197–208, 1956.

Brown, George. *Social Psychology.* New York: The Free Press, 1966.

Buckley, Walter. *Sociology and Modern Systems Theory.* Englewood Cliffs, N. J.: Prentice-Hall, 1967.

Dahrendorf, R. *Class and Class Conflict in Industrial Society.* Stanford: Stanford University Press, 1959.

DeGreene, Kenyon B. "Systems Analysis Techniques." In *Systems Psychology,* edited by Kenyon B. DeGreene. New York: McGraw-Hill, 1970.

———. *Sociotechnical Systems: Factors in Analysis, Design, and Management.* Englewood Cliffs, N. J.: Prentice-Hall, 1973.

Deutsch, Karl. "Communication Theory and Social Science." *American Journal of Orthopsychiatry* 22:469–83, 1952.

Fishburn, Peter C. *Decision and Value Theory.* New York: John Wiley & Sons, 1964.

Forrester, Jay W. *World Dynamics.* Cambridge, Mass.: Wright-Allen Press, 1971.

Fortes, Meyer. *The Dynamics of Clanship Among the Tallensi.* London: Oxford University Press, 1945.

Gastil, R. P. "A General Framework for Social Science." *Policy Sciences* 3:385–403, 1972.

Geertz, Clifford. *Agricultural Involution.* Berkeley, CA: University of California Press, 1963.

Gluckman, M. "The Utility of the Equilibrium Model in the Study of Social Change." *American Anthropologist* 70(2): 219–237, 1968.

Goldschmidt, W. R. *Man's Way.* New York: Holt, Rinehart and Winston, 1959.

Goodenough, W. H. *Cooperation in Change.* New York: Russell Sage Foundation, 1963.

Gouldner, Alvin. "Organizational Analysis" in *Sociology Today.* New York: Basic Books, 1963.

Graebner, Fritz. *Methode der Ethnologie,* Heidelberg: C. Winter, 1911.

Gubbins, J. H. *The Making of Modern Japan.* London: Lippincott, 1922.

Hallowell, A. I. Acculturation Processes and Personality Changes. In *Personality in Nature, Society and Culture,* edited by C. Kluckhohn and H. A. Murray. New York: Alfred A. Knopf, 1948.

Harris, Marvin. *The Rise of Cultural Theory.* New York: Thomas Y. Crowell Co., 1968.

———. *Culture, Man and Nature.* New York: Thomas Y. Crowell Co., 1971.

Honjo, Eijiro. *Social and Economic History of Japan.* Kyoto: Institute for Research in Economic History of Japan, 1935.

Kiernan, E. W. G. *British Diplomacy in China 1880–1885.* New York: Octogon Books, 1970.

Kluckhohn, Clyde. "The Influence of Psychiatry on Anthropology During the Past One Hundred Years." In *One Hundred Years of American*

Psychiatry, edited by J. K. Hall et al. New York: Columbia University Press, 1944.

Kluckhohn, Clyde and O. H. Mowrer. "Culture and Personality: A Conceptual Framework." *American Anthropologist* 46: 1–29, 1944.

Kobayashi, U. *Military Industries of Japan.* New York: Oxford University Press, 1922.

Kroeber, A. L. *Configurations of Cultural Growth.* Berkeley, CA: University of California Press, 1944.

______. *The Nature of Culture.* Chicago: University of Chicago Press, 1952.

Lasswell, H. *Power and Personality.* New York: W. W. Norton, 1948.

Latourette, K. S. *The Development of Japan.* New York: Macmillan, 1938.

Leach, E. R. *Political Systems of Highland Burma.* (orig. 1954) Boston: Beacon Press, 1965.

Levi-Strauss, Claude. *Structural Anthropology.* New York: Basic Books, 1963.

Linton, Ralph. *The Individual and His Society.* Edited by A. Kardiner. New York: Columbia University Press, 1939.

______. "Nativistic Movements." *American Anthropologist* 45:230–240, 1943.

Loewe, Michael. *Imperial China.* New York: Frederick A. Praeger, 1966.

Martino, Joseph P. *Technological Forecasting for Decision Making.* New York: American Elsevier Publishing Co., 1972.

Marx, K. *Capital,* 2 vols. (orig. 1867) New York: International Publishers, 1967.

Maruyama, Magorah. "The Second Cybernetics: Deviation-Amplifying Mutual Causal Processes." *American Scientist* 51:164–79, 1963.

Merton, Robert K. *Social Theory and Social Structure,* 2nd ed. New York: The Free Press, 1968.

Mesarovic, Mihajlo D., ed. *Views on General Systems Theory: Proceedings.* New York: John Wiley & Sons, 1964.

Morgan, L. H. *Ancient Society.* New York: Holt, Rinehart and Winston, 1877.

Mumford, L. *Technics and Civilization.* New York: Colliers, 1954.

Murdock, George P. *Social Structure.* New York: Macmillan, 1949.

———. "Ethnographic Atlas: A Summary." *Ethnology* 6:109–236, 1967.
———. "Anthropology's Mythology." *Proceedings of the Royal Anthropological Institute of Great Britain and Ireland,* pp. 17–23, 1971.
Nitobe, I., ed. *Western Influences in Modern Japan.* Chicago: Institute of Pacific Relations, 1930.
Norman, E. H. *Japan's Emergence as a Modern State.* New York: Institute of Pacific Relations, 1940.
Ogburn, W. F. *Social Change.* New York: Viking Press, 1950.
Parsons, H. *Man Machine Systems Experiments.* Baltimore: Johns Hopkins, 1972.
Parsons, Talcott. *The Social System.* New York: The Free Press, 1954.
Parsons, Talcott and Edward Shils. *Toward a General Theory of Action.* Cambridge, Mass.: Harvard University Press, 1951.
———. *Toward a General Theory of Action.* Cambridge, Mass.: Harvard University Press, 1954.
Parsons, Talcott and R. Bales. *Family, Socialization and Interaction Process.* New York: The Free Press, 1955.
Radcliffe-Brown, A. R. *Structure and Function in Primitive Society.* New York: The Free Press, 1952.
Rau, J. G. *Optimization and Probability in Systems Engineering.* New York: Van Nostrand, Reinhold, 1970.
Rigney, J. W. "Maintainability: Psychological Factors in Design." In *Systems Psychology,* edited by Kenyon B. DeGreene. New York: McGraw-Hill, 1970.
Sahlins, M. D. and E. R. Service, eds. *Evolution and Culture.* Ann Arbor: University of Michigan Press, 1960.
Samson, G. B. *Japan, A Short Cultural History.* New York. Appleton-Century-Crofts, 1931.
Senders, G. A. "The Estimation of Operator Workload in Complex Systems." In *Systems Psychology,* edited by Kenyon B. DeGreene. New York: McGraw-Hill, 1970.
Shinners, S. M. *Techniques of Systems Engineering.* New York: McGraw-Hill, 1967.
Smith, G. E. *In the Beginning: The Origin of Civi-*

lization. New York: William Morrow, 1928.
Sorokin, P. A. *Social and Cultural Dynamics.* New York: American Book Co., 1937–41.
Spencer, H. *Principles of Sociology,* 3 vols. New York: D. Appleton and Co., 1876–1896. (Portions of Vols. I [1876] and II [1882] are reprinted in Carniero 1967.)
Spengler, O. *The Decline of the West.* New York: Alfred Knopf, 1927.
Spiro, Melford E. "A Typology of Social Structure and the Patterning of Social Institutions: A Cross-Cultural Study." *American Anthropologist* 67:1097–1119, 1965.
Ssu-yii, Teng and John K. Fairbank. *China's Response to the West: A Documentary Survey,* 1839–1923. Cambridge, Mass: Harvard University Press, 1954.
Steward, J. H. *Theory of Culture Change.* Urbana, Ill.: University of Illinois Press, 1955.
Textor, Robert B. *Cross-Cultural Summary.* New Haven, Conn.: Human Relations Area File, 1967.
Thierauf, Robert J. *Decision Making Through Operations Research.* New York. John Wiley & Sons, 1970.
Toynbee, Arnold. *A Study of History.* New York: Oxford Press, 1947.
Turner, Victor. *Dreams, Fields and Metaphors.* Ithaca, N.Y.: Cornell University Press, 1974.
Tylor, E. B. *Primitive Culture,* 2 vols. (orig. 1871) New York: Holt, Rinehart and Winston, 1874.
______. *Anthropology.* (orig. 1881) New York: Appleton-Century-Crofts, 1897.
Vayda, A. P. "An Ecological Approach to Cultural Anthropology." *Bucknell Review* 17:112–119, 1969.
Wallace, A. F. C. "Revitilization Movements." *American Anthropologist* 58:305–317, 1956.
Weber, Max. *From Max Weber: Essays in Sociology,* Trans. and ed. by H. H. Gerth and C. Wright Mills. New York: Oxford, 1946.
White, L. *The Science of Culture.* New York: Grove Press, 1949.
______. *The Evolution of Culture.* New York: McGraw-Hill, 1959.
Wiener, Norbert. *The Human Use of Human Beings: Cybernetics and Society.* Garden City, N. Y.: Doubleday, 1954.

index